AF654631

ELECTRONIC DEVICES AND CIRCUITS

ELECTRONIC DEVICES AND CIRCUITS

Er. N.B.L. Mathur

ANMOL PUBLICATIONS PVT. LTD.
NEW DELHI-110 002 (INDIA)

ANMOL PUBLICATIONS PVT. LTD.
Regd. Office: 4360/4, Ansari Road, Daryaganj,
New Delhi-110002 (India)
Tel.: 23278000, 23261597, 23286875, 23255577
Fax: 91-11-23280289
Email: anmolpub@gmail.com
Visit us at: www.anmolpublications.com

Branch Office: No. 1015, Ist Main Road, BSK IIIrd Stage
IIIrd Phase, IIIrd Block, Bangalore-560 085 (India)
Tel.: 080-41723429 • Fax: 080-26723604
Email: anmolpublicationsbangalore@gmail.com

Electronic Devices and Circuits

© Reserved

First Edition, 2010

ISBN 978-81-261-4380-1

PRINTED IN INDIA

Printed at Mehra Offset Press, Delhi.

Contents

Preface

The book Electronic Devices and Circuits is intended to provide students with a thorough introduction to the operation of all important electronic devices in use today. With clear explanations, it shows how each device is used in appropriate electronic circuits and how simple it is to design such circuits. Numerous practical circuit design examples are offered using currently-available devices and standard-value components. Beginning with an introduction to semiconductors and pn-junction theory, the characteristics and parameters of each device are discussed in detail. Circuit applications are then treated, and the design and analysis aspects of circuits involving the device are investigated.

Designed specifically for undergraduate students of Electronics and Electrical Engineering and its related disciplines, this book offers an excellent coverage of all essential topics and provides a solid foundation for analyzing electronic circuits.

The students are thoroughly introduced to the full spectrum of fundamental topics beginning with the theory of semiconductors and p-n junction behaviour. The devices treated include diodes, transistors-BJTs, JFETs and MOSFETs- and thyristors. The circuitry covered comprises small signal (ac), power amplifiers, oscillators, and operational amplifiers including many important applications of those versatile devices.

Author

Chapter 1

Crystal Properties and Charge Carriers in Semiconductors

SEMICONDUCTOR MATERIALS

A single crystal of either an elemental (e.g., silicon) or compound (e.g., gallium arsenide) semiconductor forms the basis of almost all semiconductor devices. The ability to control the electronic and opto-electronic properties of these materials is based on an understanding of their structure. In addition, the metals and many of the insulators employed within a microelectronic device are also crystalline. Each of the semiconducting phases of the group IV (14) elements, C (diamond), Si, Ge, and α-Sn, adopt the diamond cubic structure (Figure below). Their lattice constants (*a*, Å) and densities (ρ, g/cm^3) are given in Table.

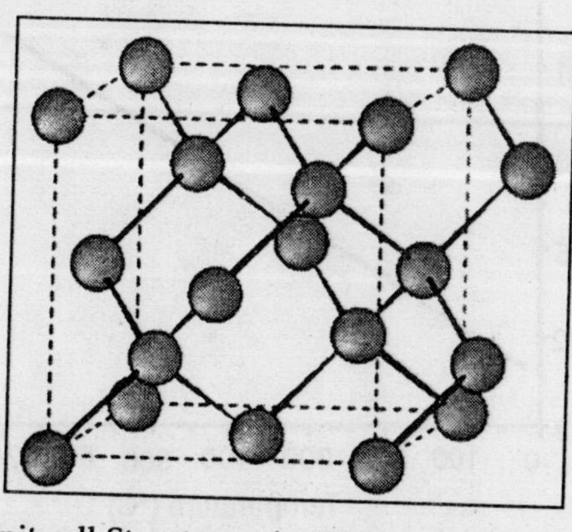

Fig. Unit cell Structure of a Diamond Cubic Lattice

Table. Lattice Parameters and Densities for the Diamond

Element	Lattice parameter, *a* (Å)	Density (g/cm³)
Carbon (diamond)	3.56683(1)	3.51525
Silicon.	5.4310201(3)	2.319002
Germanium	5.657906(1)	5.3234
Tin (α-Sn)	6.4892(1)	7.285

As would be expected the lattice parameter increase in the order C < Si < Ge < α-Sn. Silicon and germanium form a continuous series of solid solutions with gradually varying parameters.

It is worth noting the high degree of accuracy that the lattice parameters are known for high purity crystals of these elements.

In addition, it is important to note the temperature at which structural measurements are made, since the lattice parameters are temperature dependent (Figure below).

The lattice constant (*a*), in Å, for high purity silicon may be calculated for any temperature (T) over the temperature range 293 - 1073 K by the formula shown below.

$$a_T = 5.4304 + 1.8138 \times 10^{-5} (T - 298.15\text{ K}) + 1.542 \times 10^{-9} (T - 298.15\text{ K})$$

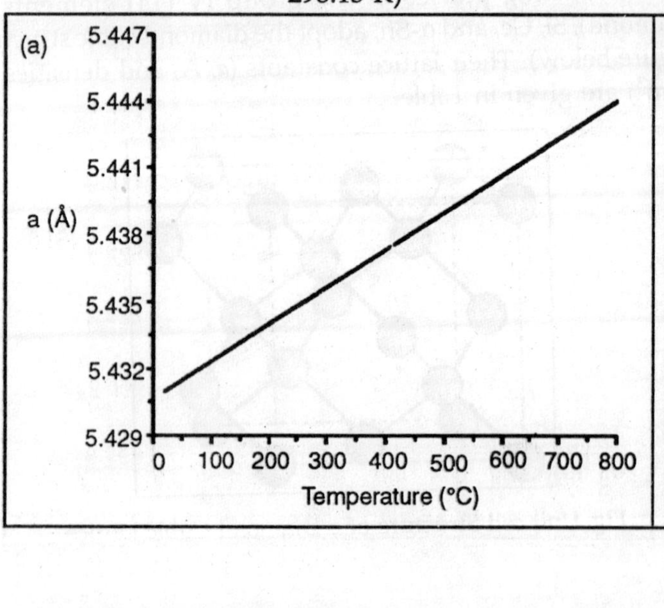

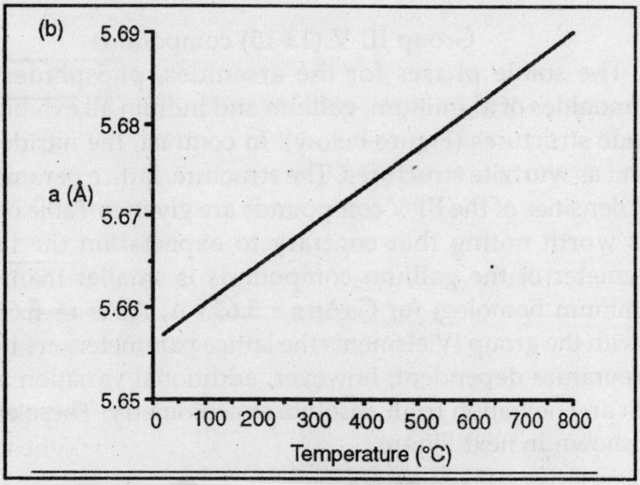

Fig. Temperature Dependence of the Lattice Parameter for (a) Si and (b) Ge.

Even though the diamond cubic forms of Si and Ge are the only forms of direct interest to semiconductor devices, each exists in numerous crystalline high pressure and meta-stable forms. These are described along with their interconversions in Table.

Table. High Pressure and Metastable Phases of Silicon and Germanium.

Phase	Structure	Remarks
Si I	diamond cubic	stable at normal pressure
Si II	grey tin structure	formed from Si I or Si V above 14GPa
Si III	cubic	metastable, formed from Si II above 10 GPa
Si IV	hexagonal	
Si V	unidentified	stable above 34 GPa, formed from Si II above 16 GPa
Si VI	hexagonal close packed	stable above 45 GPa
Ge I	diamond cubic	low-pressure phase
Ge II	β-tin structure	formed from Ge I above 10 GPa
Ge III	tetragonal	formed by quenching Ge II at low pressure
Ge IV	body centered cubic	formed by quenching Ge II to 1 atm at 200 K

Group III-V (13-15) compounds

The stable phases for the arsenides, phosphides and antimonides of aluminum, gallium and indium all exhibit zinc blende structures (Figure below). In contrast, the nitrides are found as wurtzite structures. The structure, lattice parameters, and densities of the III-V compounds are given in Table below. It is worth noting that contrary to expectation the lattice parameter of the gallium compounds is smaller than their aluminum homolog; for GaAs a = 5.653 Å; AlAs a = 5.660 Å. As with the group IV elements the lattice parameters are highly temperature dependent; however, additional variation arises from any deviation from absolute stoichiometry. These effects are shown in next Figure.

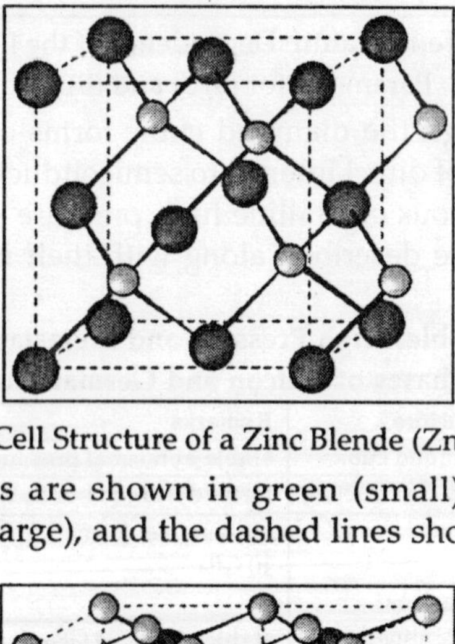

Fig. Unit Cell Structure of a Zinc Blende (ZnS) Lattice.

Zinc atoms are shown in green (small), sulfur atoms shown in red (large), and the dashed lines show the unit cell.

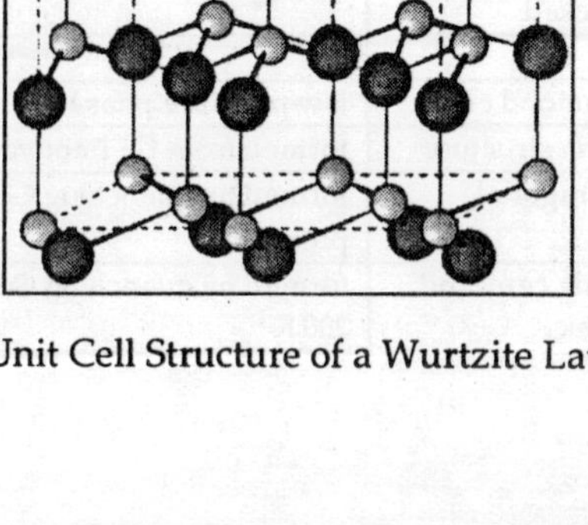

Fig. Unit Cell Structure of a Wurtzite Lattice.

Zinc atoms are shown in green (small), sulfur atoms shown in red (large), and the dashed lines show the unit cell.

Table. Lattice Parameters and Densities (Measured at 298 K) for the III-V (13-15) Compound Semiconductors.

Compound	Structure	Lattice parameter (Å)	Density (g/cm³)
AlN	wurtzite	a = 3.11(1), c = 4.98(1)	3.255
AlP	zinc blende	a = 5.4635(4)	2.40(1)
AlAs	zinc blende	a = 5.660	3.760
AlSb	zinc blende	a = 6.1355(1)	4.26
GaN	wurtzite	a = 3.190, c = 5.187	
GaP	zinc blende	a = 5.4505(2)	4.138
GaAs	zinc blende	a = 5.65325(2)	5.3176(3)
InN	wurtzite	a = 3.5446, c = 5.7034	6.81
InP	zinc blende	a = 5.868(1)	4.81
InAs	zinc blende	a = 6.0583	5.667
InSb	zinc blende	a = 6.47937	5.7747(4)

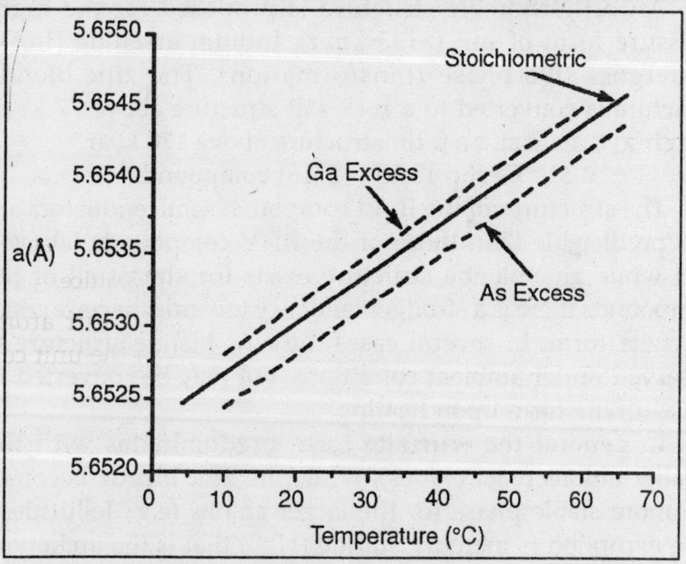

Fig. Temperature Dependence of the Lattice Parameter

The homogeneity of structures of alloys for a wide range of solid solutions to be formed between III-V compounds in almost any combination. Two classes of ternary alloys are formed: III_x–III_{1-x}–V (e.g., Al_x–Ga_{1-x}–As) and III–V_{1-x}–V_x (e.g.,

Ga–As_{1-x}–P_x). While quaternary alloys of the type III_x–III_{1-x}–V_y–V_{1-y} allow for the growth of materials with similar lattice parameters, but a broad range of band gaps. A very important ternary alloy, especially in optoelectronic applications, is Al_x–Ga_{1-x}–As and its lattice parameter (*a*) is directly related to the composition (x).

$$a = 5.6533 + 0.0078\,x$$

Not all of the III-V compounds have well characterized high-pressure phases. however, in each case where a high-pressure phase is observed the coordination number of both the group III and group V element increases from four to six. Thus, AlP undergoes a zinc blende to rock salt transformation at high pressure above 170 kbar, while AlSb and GaAs form orthorhombic distorted rock salt structures above 77 and 172 kbar, respectively.

An orthorhombic structure is proposed for the high-pressure form of InP (>133 kbar). Indium arsenide (InAs) undergoes two-phase transformations. The zinc blende structure is converted to a rock salt structure above 77 kbar, which in turn forms a β-tin structure above 170 kbar.

Group II–VI (12–16) compounds

The structures of the II–VI compound semiconductors are less predictable than those of the III–V compounds (above), and while zinc blende structure exists for almost all of the compounds there is a stronger tendency towards the hexagonal wurtzite form. In several cases the zinc blende structure is observed under ambient conditions, but may be converted to the wurtzite form upon heating.

In general the wurtzite form predominates with the smaller anions (e.g., oxides), while the zinc blende becomes the more stable phase for the larger anions (e.g., tellurides). One exception is mercury sulfide (HgS) that is the archetype for the trigonal cinnabar phase. Table listed below shows the stable phase of the chalcogenides of zinc, cadmium and mercury, along with their high temperature phases where applicable. Solid solutions of the II–VI compounds are not as easily formed as for the III-V compounds; however, two important examples are ZnS_xSe_{1-x} and $Cd_xHg_{1-x}Te$.

Table. Lattice Parameters and Densities (Measured at 298 K) for the II-VI (12-16) Compound Semiconductors.

Compound	Structure	Lattice parameter (Å)	Density (g/cm^3)
ZnS	zinc blende	a = 5.410	4.075
	wurtzite	a = 3.822, c = 6.260	4.087
ZnSe	Zinc blende	a = 5.668	5.27
ZnTe	Zinc blende	a = 6.10	5.636
CdS	wurtzite	a = 4.136, c = 6.714	4.82
CdSe	wurtzite	a = 4.300, c = 7.011	5.81
CdTe	Zinc blende	a = 6.482	5.87
HgS	cinnabar	a = 4.149, c = 9.495	
	Zinc blende	a = 5.851	7.73
HgSe	Zinc blende	a = 6.085	8.25
HgTe	Zinc blende	a = 6.46	8.07

The zinc chalcogenides all transform to a cesium chloride structure under high pressures, while the cadmium compounds all form rock salt high-pressure phases (Figure). Mercury selenide (HgSe) and mercury telluride (HgTe) convert to the mercury sulfide archetype structure, cinnabar, at high pressure.

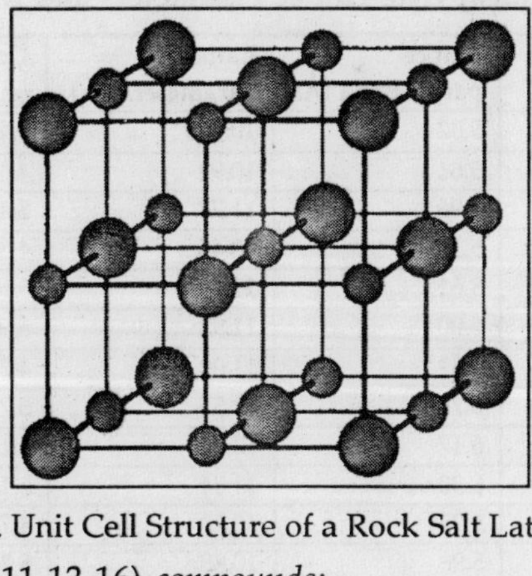

Fig. Unit Cell Structure of a Rock Salt Lattice.

I-III-VI_2 (11-13-16) *compounds:*

Nearly all I-III-VI_2 compounds at room temperature adopt the chalcopyrite structure (Figure). The cell constants and densities are given in Table below. Although there are few reports of high temperature or high-pressure phases,

$AgInS_2$ has been shown to exist as a high temperature orthorhombic polymorph ($a = 6.954$, $b = 8.264$, and $c = 6.683$ Å), and $AgInTe_2$ forms a cubic phase at high pressures.

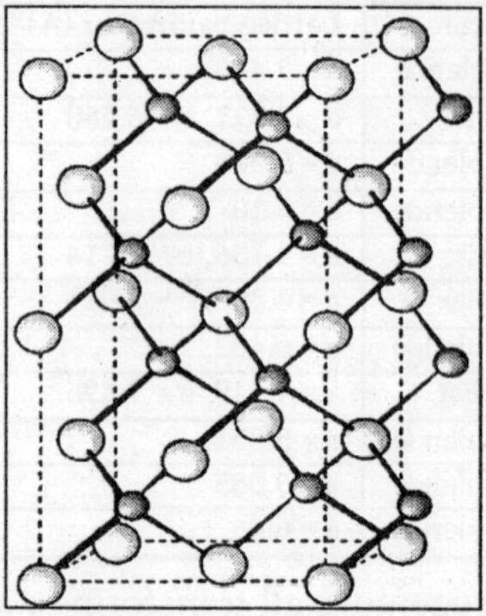

Fig. 7 Unit Cell Structure of a Chalcopyrite Lattice.

Copper atoms are shown in blue, iron atoms are shown in green and sulfur atoms are shown in yellow. The dashed lines show the unit cell.

Table. Chalcopyrite Lattice Parameters and Densities

Compound	Lattice parameter a (Å)	Lattice parameter c (Å)	Density (g.cm³)
$CuAlS_2$	5.32	10.430	3.45
$CuAlSe_2$	5.61	10.92	4.69
$CuAlTe_2$	5.96	11.77	5.47
$CuGaS_2$	5.35	10.46	4.38
$CuGaSe_2$	5.61	11.00	5.57
$CuGaTe_2$	6.00	11.93	5.95
$CuInS_2$	5.52	11.08	4.74
$CuInSe_2$	5.78	11.55	5.77
$CuInTe_2$	6.17	12.34	6.10
$AgAlS_2$	6.30	11.84	6.15
$AgGaS_2$	5.75	10.29	4.70
$AgGaSe_2$	5.98	10.88	5.70
$AgGaTe_2$	6.29	11.95	6.08
$AgInS_2$	5.82	11.17	4.97
$AgInSe_2$	6.095	11.69	5.82
$AgInTe_2$	6.43	12.59	6.96

Of the I-III-VI_2 compounds, the copper indium chalcogenides ($CuInE_2$) are certainly the most studied for their application in solar cells.

One of the advantages of the copper indium chalcogen ide compounds is the formation of solid solutions (alloys) of the formula $CuInE_{2-x}E'_x$, where the composition variable (x) varies from 0 to 2. The $CuInS_{2-x}Se_x$ and $CuInSe_{2-x}Te_x$ systems have also been examined, as has the $CuGa_yIn_{1-y}S_{2-x}Se_x$ quaternary system.

As would be expected from a consideration of the relative ionic radii of the chalcogenides the lattice parameters of the $CuInS_{2-x}Se_x$ alloy should increase with increased selenium content. Vergard's law requires the lattice constant for a linear solution of two semiconductors to vary linearly with composition (e.g., as is observed for $Al_xGa_{1-x}As$), however, the variation of the tetragonal lattice constants (*a* and *c*) with composition for $CuInS_{2-x}S_x$ are best described by the parabolic relationships.

$$a = 5.532 + 0.0801x + 0.0260\,x^2$$
$$c = 11.156 + 0.1204\,x + 0.0611\,x^2$$

A similar relationship is observed for the $CuInSe_{2-x}Te_x$ alloys.

$$a = 5.783 + 0.1560\,x + 0.0212\,x^2$$
$$c = 11.628 + 0.3340\,x + 0.0277\,x^2$$

The large difference in ionic radii between S and Te (0.37 Å) prevents formation of solid solutions in the $CuInS_{2-x}Te_x$ system, however, the single alloy $CuInS_{1.5}Te_{0.5}$ has been reported.

Orientation Effects

Once single crystals of high purity silicon or gallium arsenide are produced they are cut into wafers such that the exposed face of these wafers is either the crystallographic {100} or {111} planes.

The relative structure of these surfaces are important with respect to oxidation, etching and thin film growth. These processes are orientation-sensitive; that is, they depend on the direction in which the crystal slice is cut.

Atom Density and Dangling Bonds

The principle planes in a crystal may be differentiated in a number of ways, however, the atom and/or bond density are useful in predicting much of the chemistry of semiconductor surfaces. Since both silicon and gallium arsenide are *fcc*structures and the {100} and {111} are the only technologically relevant surfaces, discussions will be limited to *fcc* {100} and {111}.

The atom density of a surface may be defined as the number of atoms per unit area. Figure 8 shows a schematic view of the {111} and {100} planes in a *fcc* lattice. The {111} plane consists of a hexagonal close packed array in which the crystal directions within the plane are oriented at 60° to each other. The hexagonal packing and the orientation of the crystal directions are indicated in Figure as an overlaid hexagon.

Given the intra-planar inter-atomic distance may be defined as a function of the lattice parameter, the area of this hexagon may be readily calculated. For example in the case of silicon, the hexagon has an area of 38.30 $Å^2$. The number of atoms within the hexagon is three: the atom in the centre plus 1/3 of each of the six atoms at the vertices of the hexagon (each of the atoms at the hexagons vertices is shared by three other adjacent hexagons).

Thus, the atom density of the {111} plane is calculated to be 0.0783 $Å^{-2}$. Similarly, the atom density of the {100} plane may be calculated. The {100} plane consists of a square array in which the crystal directions within the plane are oriented at 90° to each other. Since the square is coincident with one of the faces of the unit cell the area of the square may be readily calculated. For example in the case of silicon, the square has an area of 29.49 $Å^2$.

The number of atoms within the square is 2: the atom in the centre plus 1/4 of each of the four atoms at the vertices of the square (each of the atoms at the corners of the square are shared by four other adjacent squares).

Thus, the atom density of the {100} plane is calculated to be 0.0678 $Å^{-2}$. While these values for the atom density are specific for silicon, their ratio is constant for all diamond cubic and zinc blende structures: {100}:{111} = 1:1.155. In general, the fewer dangling bonds the more stable a surface structure.

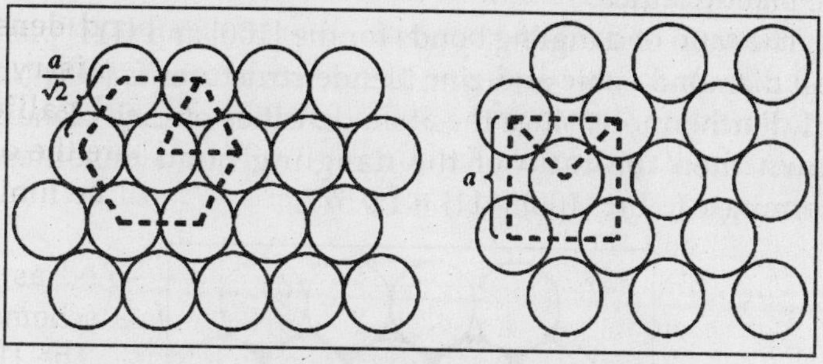

Fig. Schematic Representation of the (111) and (100) Faces of a Face Centered Cubic (fcc)

An atom inside a crystal of any material will have a coordination number (n) determined by the structure of the material. For example, all atoms within the bulk of a silicon crystal will be in a tetrahedral four-coordinate environment (n = 4). However, at the surface of a crystal the atoms will not make their full compliment of bonds.

Each atom will therefore have less nearest neighbors than an atom within the bulk of the material. The missing bonds are commonly called dangling bonds. While this description is not particularly accurate it is, however, widely employed and as such will be used herein.

The number of dangling bonds may be defined as the difference between the ideal coordination number (determined by the bulk crystal structure) and the actual coordination number as observed at the surface.

Figure below shows a section of the {111} surfaces of a diamond cubic lattice viewed perpendicular to the {111} plane. The atoms within the bulk have a coordination number of four. In contrast, the atoms at the surface are each bonded to just three other atoms, thus each surface atom has one dangling bond.

As can be seen from Figure 10, which shows the atoms at the {100} surface viewed perpendicular to the {100} plane, each atom at the surface (e.g., the atom shown in Figure below) is only coordinated to two other atoms, leaving two dangling bonds per atom. It should be noted that the same number of dangling bonds are found for the {111} and {100} planes of a zinc blende lattice.

The ratio of dangling bonds for the {100} and {111} planes of all diamond cubic and zinc blende structures is {100}:{111} = 2:1. Furthermore, since the atom densities of each plane are known then the ratio of the dangling bond densities is determined to be: {100}:{111} = 1:0.577.

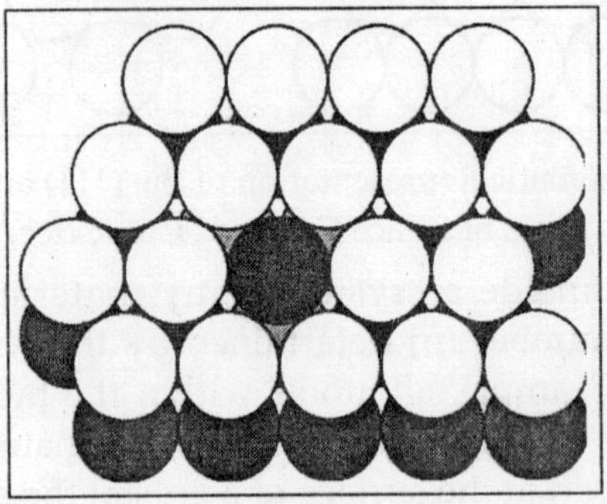

Fig. A Section of The {111} surfaces of a Diamond Cubic Lattice Viewed Perpendicular to the {111} Plane.

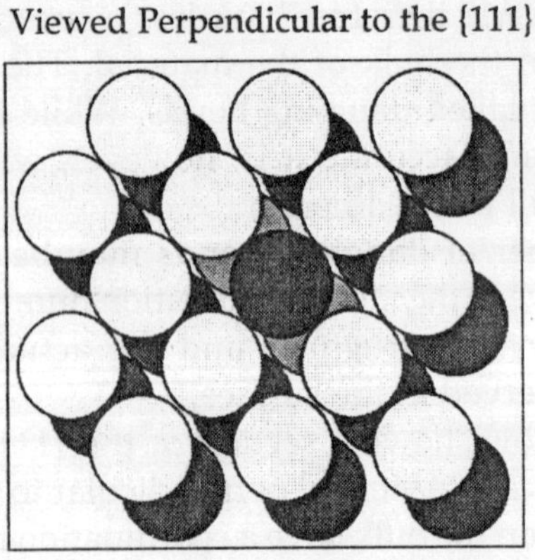

Fig. A Section of the {100} Surface of a Diamond Cubic Lattice Viewed Perpendicular to the {100} Plane.

Silicon

For silicon, the {111} planes are closer packed than the {100} planes. As a result, growth of a silicon crystal is therefore slowest in the <111> direction, since it requires laying down a close packed atomic layer upon another layer in its closest packed form. As a consequence <111> Si is the easiest to grow, and therefore the least expensive.

The dissolution or etching of a crystal is related to the number of broken bonds already present at the surface: the fewer bonds to be broken in order to remove an individual atom from a crystal, the easier it will be to dissolve the crystal. As a consequence of having only one dangling bond (requiring three bonds to be broken) etching silicon is slowest in the <111> direction. The electronic properties of a silicon wafer are also related to the number of dangling bonds.

Silicon microcircuits are generally formed on a single crystal wafer that is diced after fabrication by either sawing part way through the wafer thickness or scoring (scribing) the surface, and then physically breaking. The physical breakage of the wafer occurs along the natural cleavage planes, which in the case of silicon are the {111} planes.

Gallium Arsenide

The zinc blende lattice observed for gallium arsenide results in additional considerations over that of silicon. Although the {100} plane of GaAs is structurally similar to that of silicon, two possibilities exist: a face consisting of either all gallium atoms or all arsenic atoms. In either case the surface atoms have two dangling bonds, and the properties of the face are independent of whether the face is gallium or arsenic.

The {111} plane also has the possibility of consisting of all gallium or all arsenic. However, unlike the {100} planes there is a significant difference between the two possibilities. Figure third shows the gallium arsenide structure represented by two interpenetrating *fcc* lattices. The [111] axis is vertical within the plane of the page. Although the structure consists of alternate layers of gallium and arsenic stacked along the [111] axis, the distance between the successive layers alternates between large and small. Assigning arsenic as the parent lattice the order of the layers in the [111] direction is As Ga-As Ga-As Ga, while in the [111] direction the layers are ordered, Ga-As-Ga As-Ga

As (Figure below). In silicon these two directions are of course identical.

The surface of a crystal would be either arsenic, with three dangling bonds, or gallium, with one dangling bond. Clearly, the latter is energetically more favorable. Thus, the (111) plane shown in Figure below is called the (111) Ga face. Conversely, the [111] plane would be either gallium, with three dangling bonds, or arsenic, with one dangling bond. Again, the latter is energetically more favorable and the [111] plane is therefore called the (111) As face.

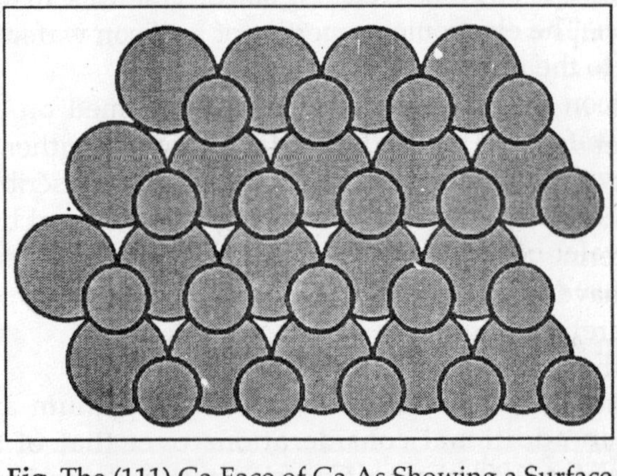

Fig. The (111) Ga Face of Ga As Showing a Surface layer of Gallium Atoms.

CRYSTAL LATTICE STRUCTURE

The arrangement of atoms, ions, or molecules in a crystal. Crystals are solids having, in all three dimensions of space, a regular repeating internal unit of structure.

Crystals have been studied using x-rays, which excite signals from the atoms. The signals are of different strengths and depend on the electron density distribution about atomic cores. Light atoms give weaker signals and hydrogen is invisible to x-rays.

However, the mutual atomic arrangements that are called crystal structures can be derived once the chemical formulas and physical densities of solids are known, based on the knowledge that atomic positions are not arbitrary but are dictated by crystal symmetry, and that the diffraction signals received are the result of systematic constructive interference between the scatterers within the regularly repeating internal unit of pattern.

Crystals are defined in terms of space, population, and mutual arrangement. Crystal space is represented as an indefinitely extended lattice of periodically repeating points. The periodicity of the lattice is defined by the lengths and mutual orientations of three lattice vectors that enclose the pattern.

Population is defined as the total number and kind of fundamental units of structure that form the pattern. The order and periodicity of crystals must extend to about 100 nanometers in all three dimensions of space to give the sharply defined diffraction signals required for mapping structural details by x-rays.

Intermediate states of order are seen in liquid crystals, which have long molecules as fundamental units of structure. These are arranged with their lengths parallel to each other, but without periodicity, in the nematic state. In the smectic state there is orientation in equally spaced planes but no sideways periodicity, like traffic moving freely on a multilane highway.

In reality crystal space is not, in general, perfect. The growth process is characterized by constraints and turbulences, and by the dynamic interaction between the crystal and its environment.

The process is reflected within the structures formed as an assemblage of atoms is collected and made relatively immobile by releasing the energy known as the heat of crystallization.

The resulting crystal lattices resemble a mosaic of slightly misaligned adjacent regions. This is fortunate for research in x-ray crystallography. Perfect alignment would result in subtraction of energy by interference with the primary beam, due to a 180° phase reversal of the reflected beam (primary extinction). Internally diffracted beams would also be attenuated by internal reflection from regions above them (secondary extinction).

Each of the spatially misaligned mosaic blocks of a single crystal is assumed to maintain lattice periodicity within it. This assumption is confirmed by the sharp diffraction patterns observed.

There are some "wrong" atoms, vacant lattice sites, trapped gas atoms, and so forth, and the atomic occupants jiggle about while also vibrating cooperatively and synchronously in complex internal modes of motion.

Intricate patterns of electron exchange are enacted, and systematic changes in spin orientations can occur for an atom with a magnetic moment.

Details like these are important for understanding the relationships between structure determination on the atomic and molecular levels and the cooperative behaviour that determines bulk properties and functions.

A rectangular space lattice with two possible cells is outlined in the illustration. These have the same cell volumes but different symmetries.

Since crystallographic unit cells are completely defined by three lattice vectors, the crystal symmetry referenced to this lattice can be no higher than orthorhombic: $a \neq b \neq c$ ($OA \neq OB \neq OC$), and all angles equal to 90°.

This and a possible monoclinic cell, with the same vectors a and b (OA and OB) and one angle not equal to 90°, are outlined.

If the OAB plane is rotated and the vector a (OA) is extended to terminate at the next lattice point, then all angles differ from 90° and the crystal symmetry represented becomes triclinic.

The mutual arrangement and atom coordinates of the cell population must be such that the environment, seen from every point of the space lattice, remains the same.

Screw axes combine the rotation of an ordinary symmetry axis with a translation parallel to it and equal to a fraction of the unit distance in this direction.

If screw axes are present in crystals, it is clear that the

displacements involved are of the order of a few tenths of nanometer and that they cannot be distinguished macroscopically from ordinary symmetry axes.

The same is true for glide mirror planes, which combine the mirror image with a translation parallel to the mirror plane over a distance that is half the unit distance in the glide direction. The handedness of screw axes is a very important feature of many biological and mineral structures.

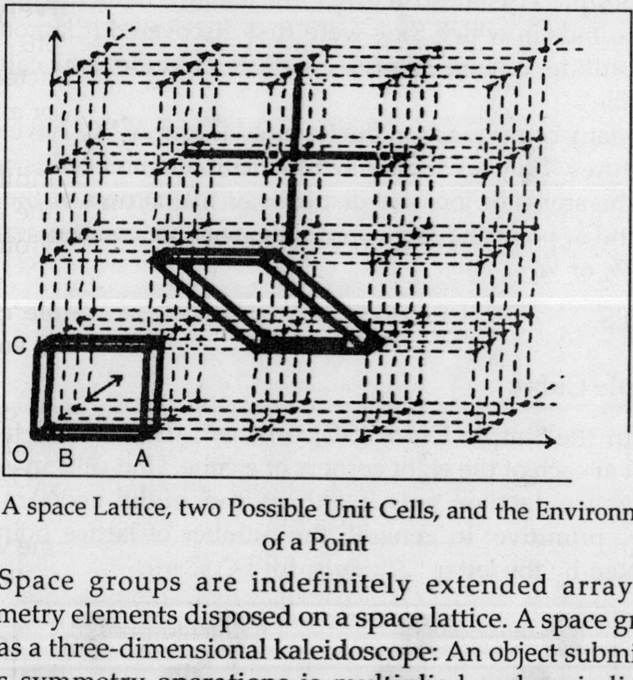

Fig. A space Lattice, two Possible Unit Cells, and the Environment of a Point

Space groups are indefinitely extended arrays of symmetry elements disposed on a space lattice. A space group acts as a three-dimensional kaleidoscope: An object submitted to its symmetry operations is multiplied and periodically repeated in such a way that it generates a number of interpenetrating identical space lattices.

In general, metallic structures are relatively simple, characterized by a close packing and a high degree of symmetry. Manganese, gallium, mercury, and one form of tungsten are exceptions.

A characteristic of metallic structures is the frequent occurrence of allotropic forms; that is, the same metal can have two or more different structures which are most frequently stable in a different temperature range. The forces which link the atoms together in metallic crystals are nondirectional. This means that each atom tends to surround itself by as many others as possible.

This results in a close packing, similar to that of spheres of equal radius, and yields three distinct systems: close-packed (face-centered) cubic, hexagonal close-packed, and body-centered cubic.

Simple crystal structures are usually named after the compounds in which they were first discovered (diamond or zinc sulfide, cesium chloride, sodium chloride, and calcium fluoride).

Many compounds of the types A^+X^- and $A^{2+}X_2^-$ have such structures. They are highly symmetrical, the unit cell is cubic, and the atoms or ions are disposed at the corners of the unit cell and at points having coordinates that are combinations of 0, 1, ½, or ¼.

TYPES

Simple Cubic

In the Simple Cubic (SC) unit cell there is one lattice point at each of the eight corners of a cube. Unit cells in which there are lattice points onlyat the eight corners are called primitive. In general, the number of lattice points is denoted by the letter "Z"; thus, for SC, Z = 1.

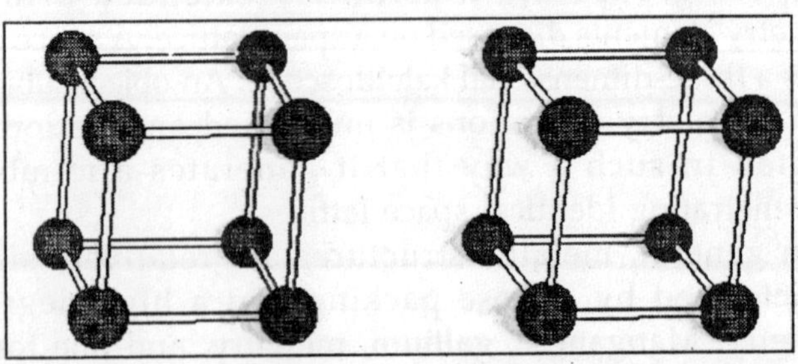

Fig.(Crossed Stereo Pair)

Let a host atom of radius r occupy each lattice point, and assume that each atom touches as many adjacent atoms as possible (in this case, there are six such contacts). Then each of the three unit cell edges is equal to the sum of two atomic radii: a = b = c = 2r.

The volume of the cell is thus:

$$V_c = 8r^3$$

In a simple cubic cell, there is one host atom wholly inside the cube, because each of the eight corner atoms contributes one eighth of an atom to the cell interior. In general, the total volume of the cell which is occupied by the host atoms is

$$V_s = {}^4/_3\pi r^3 \cdot Z.$$

The packing efficiency of a lattice is defined as the ratio V_s:V_c. Thus, for SC, the packing efficiency is about 52%.

Body Centered Cubic

In the Body Centered Cubic (BCC) unit cell there is one host atom (lattice point) at each corner of the cube and one host atom in the centre of the cube: Z = 2.

Each corner atom touches the central atom along the body diagonal of the cube, and it is easy to show by that the unit cell edge, an irrational number, is about 2.3r. Thus, the corner atoms do not touch one another.

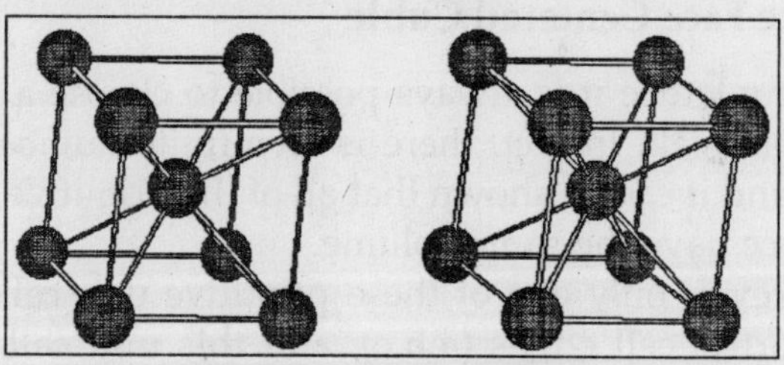

Fig. (Crossed Stereo Pair)

The packing efficiency in this lattice can be shown to be about 68%, much higher than the packing efficiency in a simple cubic lattice.

Face Centered Cubic

In the Face Centered Cubic (FCC) unit cell there is one host atom at each corner and one host atom in each face. Since each corner atom contributes one eighth of its volume to the cell interior, and each face atom contributes one half of its volume to the cell interior (and there are six faces), then $Z = {}^{1}/_{8}\cdot 8 + {}^{1}/_{2}\cdot 6 = 4$.

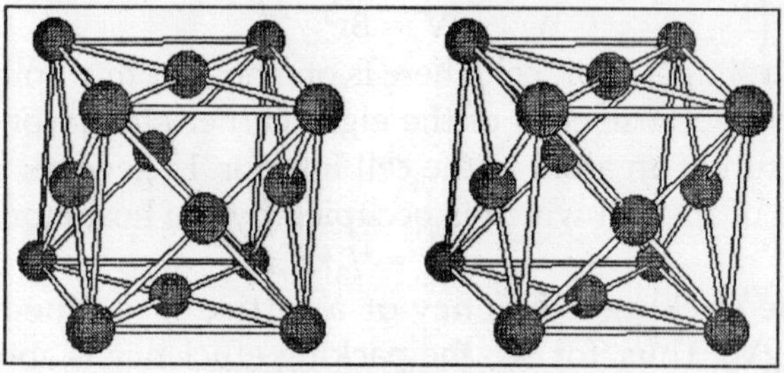

Fig. (Crossed Stereo Pair)

The corner and face atoms touch along the face diagonal, and it is easy to show that the cube edge (a) is about 2.8r. Thus, the corner atoms do nottouch one another.

The packing efficiency is about 74%. This is the maximum packing efficiency for spheres of equal radius and is call closest packing. Thus a face centered lattice of atoms is also called Cubic Closest Packing (CCP).

Primitive Face Centered Cubic

In any lattice it is always possible to choose a primitive (Z = 1) unit cell. In fact, there is an infinite number of such choices, and it can be shown that all of the primitive unit cells on a lattice have the same volume.

However, only one of these primitive unit cells has the three shortest cell edges (a,b,c), and this unit cell is called the standard reduced cell.

In a FCC lattice, the standard reduced cell is a rhombohedron, with a = b = c = 2r. The three interior angles formed between unit cell edges are called:

α (alpha, between edges b & c)

β (beta, between edges a & c)

γ (gamma, between edges a & b)

In the FCC rhombohedral standard reduced cell, it can be shown that $\alpha = \beta = \gamma = 60^o$. Note that a cube is a just a special rhombohedron, with$\alpha = \beta = \gamma = 90^o$.

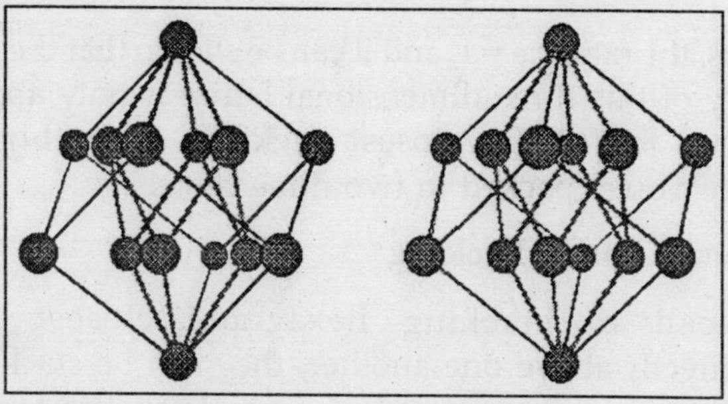

Fig. Crossed Stereo Pair

Primitive Hexagonal

In a plane, spheres of equal size are most densely packed (with the least amount of empty space) when each sphere touches six other spheres arranged in the form of a regular hexagon.

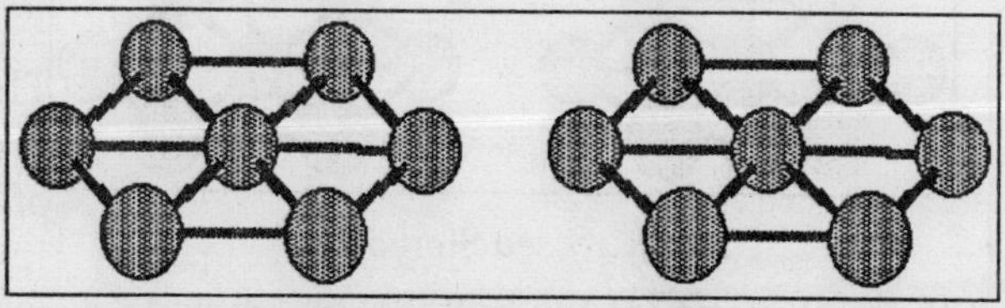

Fig. (Crossed Stereo Pair)

When two such hexagonally closest packed planes are stacked directly on top of one another, a primitive hexagonal array results.

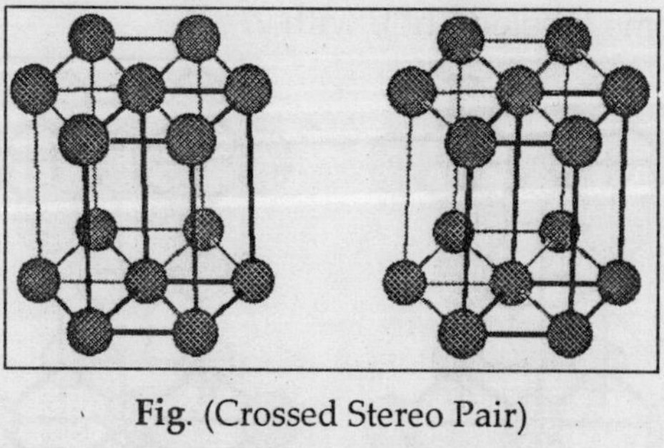

Fig. (Crossed Stereo Pair)

The primitive unit cell, outlined in black in the crossed stereo pair above, has cell edges a = b = 2r and c = 2r.

Thus, the ratio c:a = 1, and it can be shown that the packing efficiency of this three dimensional lattice is only about 60% (compared to 74% for closest packing), even though the atoms are closest packed in two dimensions.

Hexagonal Closest Packing

Instead of stacking hexagonal closest packed planes directly above one another, they can be stacked such that atoms in successive planes nestle in the triangular "grooves" of the adjacent plane. (note that there are six of these "grooves" surrounding each atom in the hexagonal plane, but only three of them can be covered by atoms in the adjacent plane).

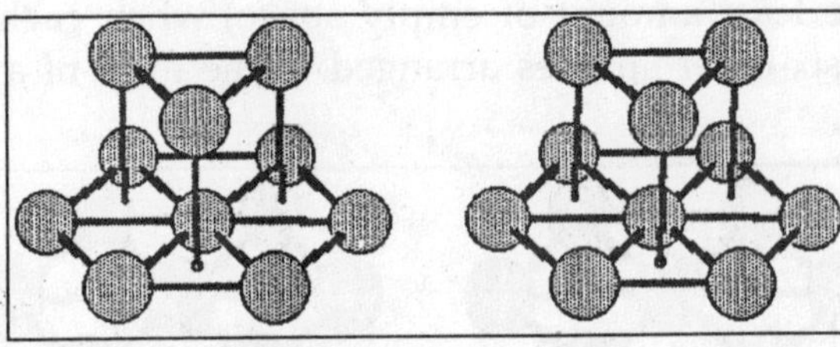

Fig. (Crossed Stereo Pair)

Let the first plane (at the bottom) be labeled "A" and the next plane above it be labeled "B". If a third hexagonal closest packed plane is stacked above B but in the "A" orientation, and succeeding planes are stacked in the repeating pattern ABABA... = (AB), a hexagonal unit cell can be chosen (using the nine atoms labeled "h"), with Z = 2.

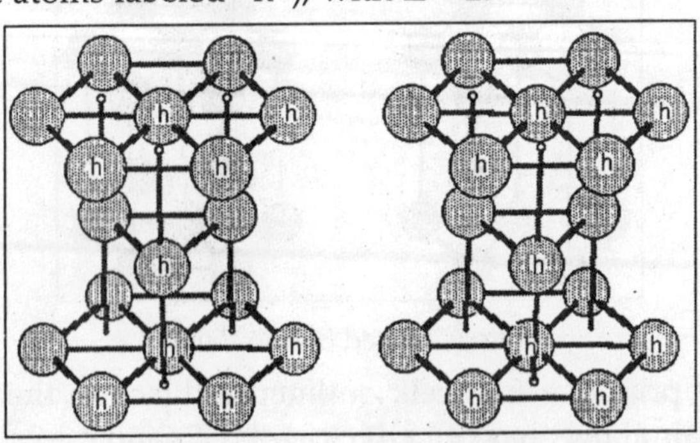

Fig. (Crossed Stereo Pair)

For the stack of hexagonally closest packed spheres of equal radius (r) described above, the interplanar spacing between adjacent planes is proportional to r.

The proportionality constant is an irrational number called the Closest Packed Interlayer Spacing, CPIS, and its value is about 1.6. Thus, the interlayer spacing is about 1.6·r (compared to 2·r for simple hexagonal stacking).

Thus, the "c" unit cell edge (in the stacking direction) has a length c = 2·CPIS·r, the ratio c:a = CPIS, and it can be shown that this lattice has a packing efficiency which is identical to the FCC lattice. Thus, the name Hexagonal Closest Packing (HCP) for this array is justified.

Coordination in HCP

Each host atom in an HCP lattice is surrounded by and touches 12 nearest neighbors, each at a distance of 2r:

- There are six atoms in the planar hexagonal array (the central A layer);
- There are three atoms in the B layer above the A layer;
- There are three atoms in the B layer below the A layer.

The six atoms in the two B layers form a trigonal prism around the central atom in the A layer; the length of this prism is 2·CPIS·r.

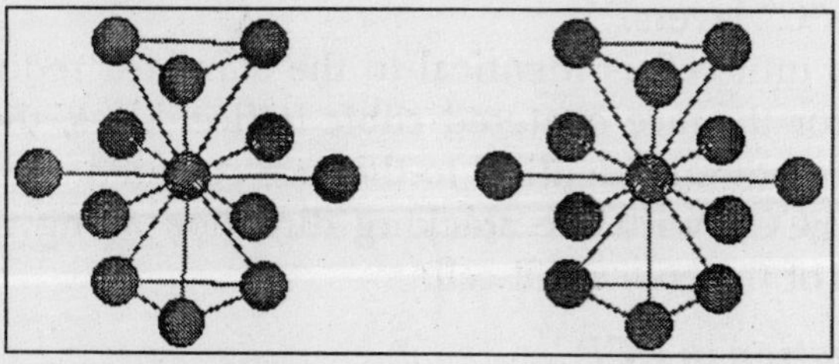

Fig. (Crossed Stereo Pair)

Cubic Closest Packing

Suppose the first two layers of hexagonal closest packed planes are stacked in "AB" fashion but the third layer is positioned so that its atoms lie over the three grooves in the A layer which were not covered by the atoms in the B layer. Then the third layer is in a different orientation from either A or B and is labeled "C".

If a fourth layer then repeats the A layer orientation, and succeeding layers repeat the pattern ABCABCA... = (ABC), the resulting unit cell is hexagonal with three host atoms (Z = 3), unit cell edge c = 3·CPIS·r and c:a = 1.5·CPIS. Note that for identical atoms in all layers, (ACB) is identical to (ABC).

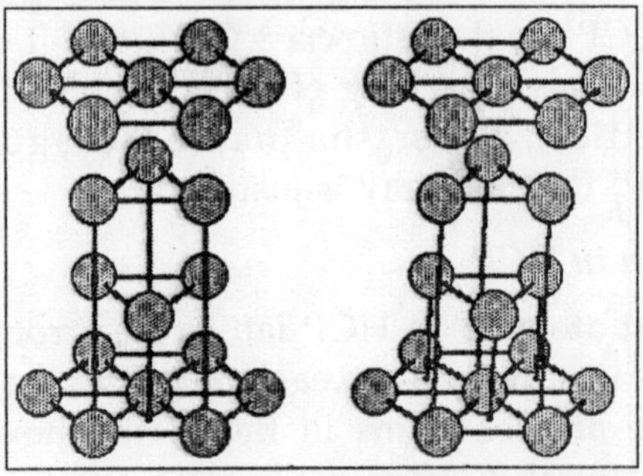

Fig. (Crossed Stereo Pair)

It can be shown that this is a closest packed structure because the three host atoms occupy 74% of the total hexagonal unit cell volume.

Furthermore, the standard reduced cell in this array is as follows: choose the two central atoms in the top and bottom "A" layers, and connect them to the six atoms shown in the "B" and "C" layers.

This unit cell is identical to the standard reduced cell chosen for the face centered cubic lattice. Thus, the (ABC) repeat structure is identical to the face centered cubic lattice (CCP = FCC), with the stacking direction along the body diagonal of the cubic unit cell.

Coordination in CCP

Each host atom in a CCP lattice is surrounded by and touches 12 nearest neighbors, each at a distance of 2r:

- There are six atoms in the central A layer;
- There are three atoms in the B layer above the A layer;

- There are three atoms in the C layer below the A layer.

The six atoms in the B and C layers form a trigonal antiprism (also known as a distorted octahedron) around the central atom; the length of this prism (the distance between the two equilateral triangular faces) is 2·CPIS·r, but the length of the long edges of the six isosceles triangular faces is slightly greater than that value.

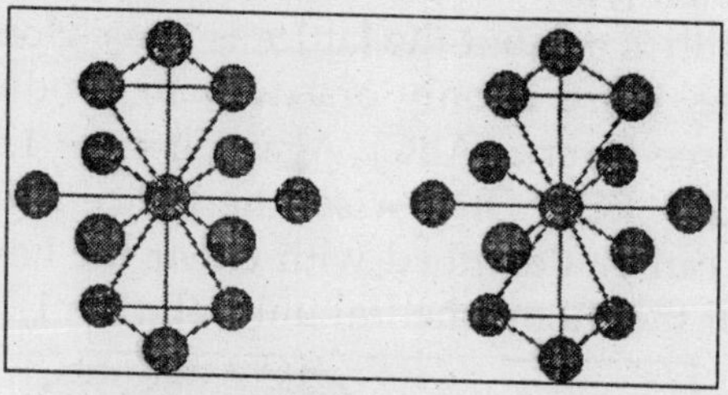

Fig. (Crossed Stereo Pair)

The Rhombohedral Lattice

In order for the (ABC) layered lattice to be closest packed, the interlayer spacing must be exactly equal to CPIS·r with c:a = 1.5·CPIS. Thus, the standard reduced cell is then the special rhombohedron found in the face centered cubic lattice.

If the c:a ratio differs from the closest packed value, then the standard reduced unit cell is still a rhombohedron ($a = b = c$ and $\alpha = \beta = \gamma$), but the cell edges need not be of length 2r and/or the inter edge angles need not be 60°. In these quasi-closest packed structures, either the hexagonal (Z = 3) cell or the primitive (Z = 1) unit cell may be used to describe this lattice, which is called, by convention, rhombohedral.

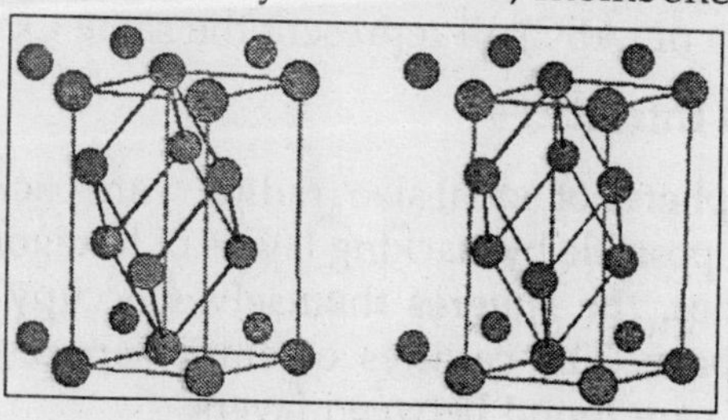

Fig. (Crossed Stereo Pair)

We see that quasi-closest packed (AB) lattices are also possible if c:a differs from CPIS.

Two and Three Layer Repeat

There is only one way to produce a repeat pattern (crystal lattice) with layers of hexagonally closest packed planes in just two orientations: (AB). If the interlayer distance is such that c:a = CPIS, then (AB) = HCP. Otherwise, the unit cell is still hexagonal with Z = 2, but the lattice is quasi-closest packed.

Likewise, there is only one way to produce a repeat pattern in three layers: (ABC). Again, if c:a = 1.5.CPIS, then (ABC) = CCP = FCC. Otherwise, the lattice is quasi-closest packed and can be described with either the hexagonal unit cell (Z = 3) or the rhombohedral unit cell (Z = 1).

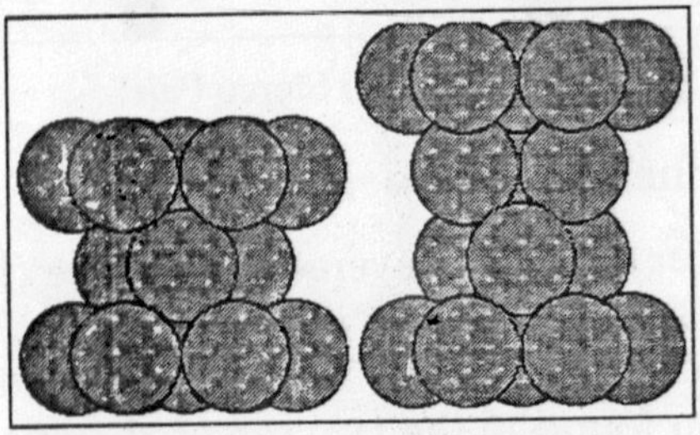

Fig. These are not Stereo Pairs

It is important to understand that apparently different permutations of the letter codes for hexagonal closest packed layers do not necessarily represent unique lattice patterns. For example, the six permutations (AB), (AC), (BA), (BC), (CA), and (CB) all represent the same HCP lattice in different orientations and choices of unit cell axes. Likewise, the six permutations of (ABC) all represent the same CCP lattice.

Tetrahedral Interstices

When spheres of equal size (radius r) are packed together as closely as possible by stacking layers of hexagonally closest packed planes, the spheres themselves occupy 74% of the available space. The cavities of empty space ("holes" or "interstices") are found between layers.

For example, consider any two successive planes in a

closest packed lattice. One atom in the A layer nestles in the triangular groove formed by three adjacent atoms in the B layer, and the four atoms touch along the edges of a regular tetrahedron.

The edges of the tetrahedron are of length 2r, and the centre of the tetrahedron is a cavity called the Tetrahedral hole (a common symbol for "tetrahedral" is T_d). If a guest sphere in that cavity is in contact with all four host spheres, its radius is about 0.22r.

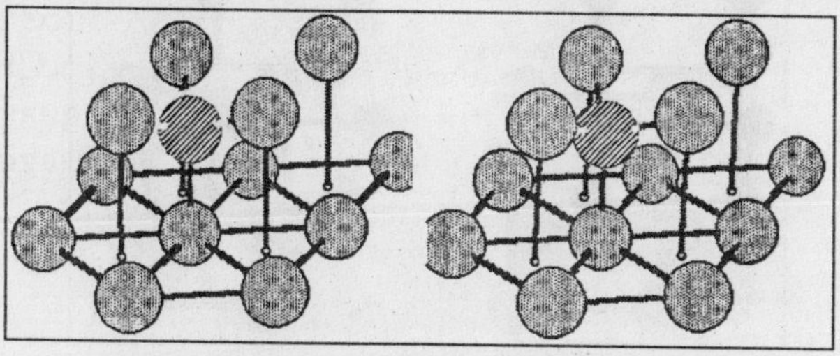

Fig. (Crossed Stereo Pair)

Octahedral Interstices

Adjacent to the T_d hole, three atoms in the B layer touch three atoms in the A layer such that a trigonal antiprismatic polyhedron (a regular octahedron) is formed. At the centre of the octahedron is a cavity called the Octahedral Hole (a common symbol for "octahedral" is O_h). A guest sphere will just fill this cavity and touch the six host spheres if its radius is approximately 0.41r. It can be shown that there are twice as many T_d holes as there are O_h holes in any closest packed bilayer.

Fig. (Crossed Stereo Pair)

EXAMPLES

Cesium Chloride

An example of a crystal structure which can be described using the ideas presented above is CsCl. Each ion resides on a separate, interpenetratingSimple Cubic lattice. The cation is in the centre of the anion unit cell and visa versa, so the two lattices have the same unit cell dimensions.

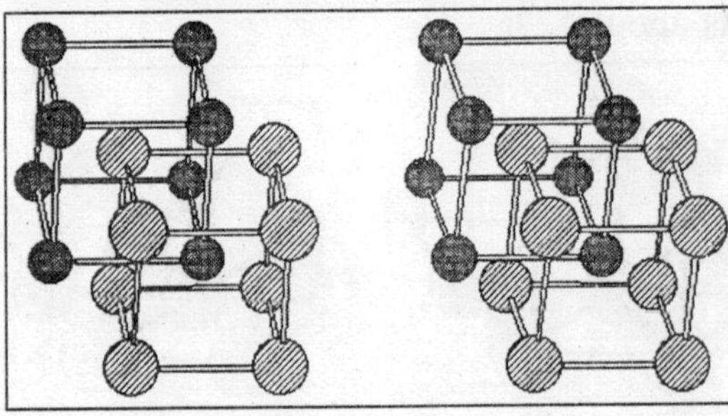

Fig. (Crossed Stereo Pair)

Note that each lattice is simple cubic, not body centered cubic. In order for a lattice to be body centered, the same chemical species (e.g., an atom or ion) would have to reside at the eight corners and the centre of the unit cell.

Sodium Chloride

In a NaCl crystal, each ion resides on a separate, interpenetrating FCC lattice, and the two lattices have the same unit cell dimensions.

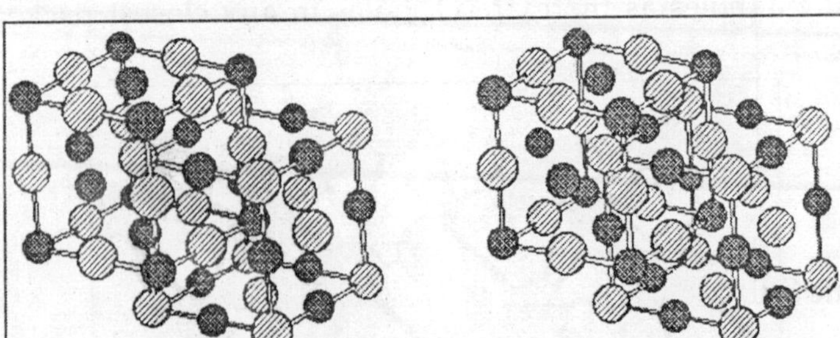

Fig. (Crossed Stereo Pair)

The Halite Structure

Sodium chloride is a specific example of a common structure type known as the "halite" structure. This lattice may also be viewed as a FCC lattice of anions (Z = 4), with smaller cations occupying all O_h holes (or vice versa). By identifying and counting all of the O_h holes, it can be shown that the number of cations wholly within the unit cell is also 4.

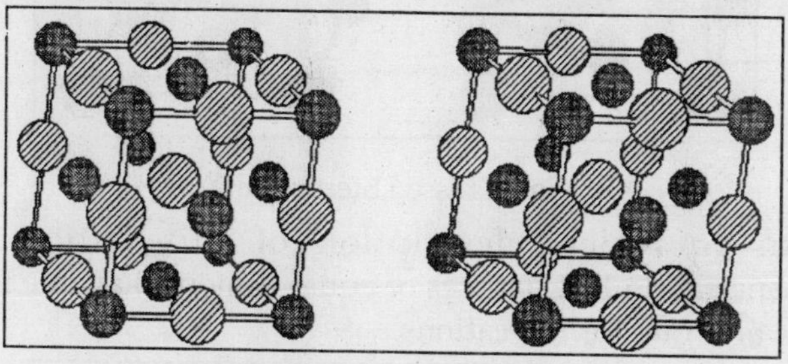

Fig. (Crossed Stereo Pair)

Calcium Fluoride

The structure of the mineral fluorite (calcium fluoride) may be viewed as a FCC lattice of cations (Z = 4), with the smaller anions occupying all of the T_d holes (Z = 8). The T_d cavities reside on a Simple Cubic lattice which is half the dimension of the CCP lattice.

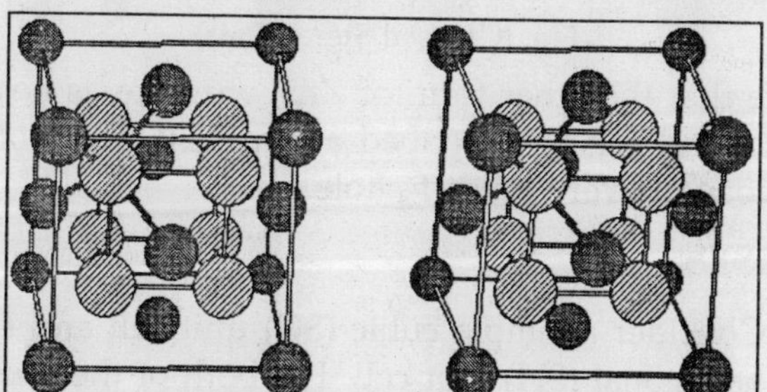

Fig. (Crossed Stereo Pair)

Zinc Sulfide

The structure of cubic ZnS (mineral name "zinc blende") may be viewed as a CCP lattice of anions (Z = 4), with the smaller cations occupyingevery other T_d hole (Z = 4) or vice versa.

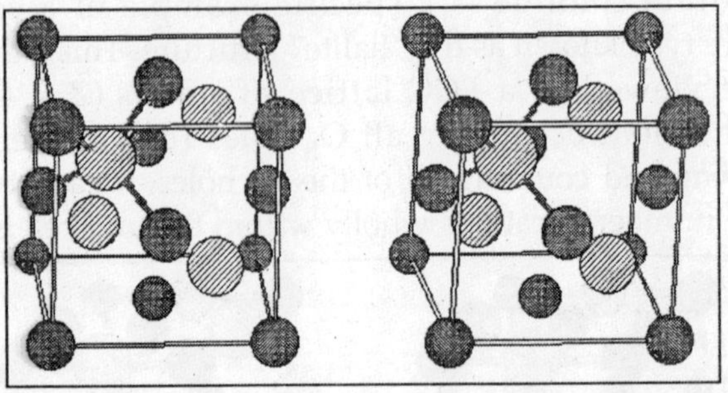

Fig. (Crossed Stereo Pair)

An equivalent description of this lattice is as interpenetrating FCC lattices of equal dimensions, one for the anions and one for the cations.

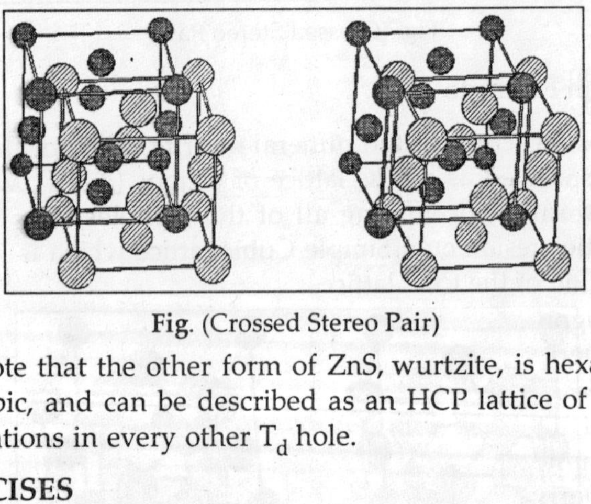

Fig. (Crossed Stereo Pair)

Note that the other form of ZnS, wurtzite, is hexagonal, not cubic, and can be described as an HCP lattice of anions with cations in every other T_d hole.

EXERCISES

- Consider a simple cubic (SC) unit cell and a simple hexagonal (SH) unit cell. For both of these cells, Z = 1 and a = b = c = 2r, where r is the radius of the host atoms which occupy the eight lattice points. Derive an exact mathematical expression for the ratio of cell volumes SC:SH, and report this irrational number to six significant figures.

- Derive an exact mathematical expression for the packing efficiency of a simple cubic unit cell, and report this irrational number to six significant figures.
- Derive an exact mathematical expression for the packing efficiency of a body centered cubic unit cell, and report this irrational number to six significant figures.
- Derive an exact mathematical expression for the packing efficiency of a face centered cubic unit cell, and report this irrational number to six significant figures.
- Prove that the inter-edge angles of the standard reduced cell in the FCC lattice are all 60°.
- Derive an exact mathematical expression for the closest packed interlayer spacing (CPIS), and report this irrational number to six significant figures. [Hint: CPIS is the height of the trigonal pyramid which is a regular tetrahedron, all six edges of which are of length 2r.)
- Derive an exact mathematical expression for the packing efficiency of a simple hexagonal cell, and report this irrational number to six significant figures.
- Derive an exact mathematical expression for the packing efficiency of a hexagonal closest packed (AB) unit cell, and report this irrational number to six significant figures.
- Derive an exact mathematical expression for the packing efficiency in the hexagonal unit cell (Z = 3) of the CCP (ABC) lattice, and report this irrational number to six significant figures.
- Derive an exact mathematical expression for the length of the longest edge of the trigonal antiprism described in the discussion of CCP Coordination, and report this irrational number to six significant figures.
- How many distinctly different closest packed lattices are possible by stacking hexagonal closest packed planes in 7-layer repeat patterns?
- Derive an exact mathematical expression for the size

of the sphere which just fills the tetrahedral hole in a closest packed bilayer, and report this irrational number to six significant figures. This radius will be a function of r, the radius of the host atoms composing the bilayer.

- Derive an exact mathematical expression for the radius of the sphere which just fills the octahedral hole in a closest packed bilayer, and report this irrational number to six significant figures. This radius will be a function of r, the radius of the host atoms composing the bilayer.
- Prove that in a closest packed lattice, there are twice as many tetrahedral as octahedral holes.

BONDING FORCES AND ENERGY BANDS IN SOLIDS

BONDING FORCES

The Coulomb forces are simple: attractive between electrons and nuclei, repulsive between electrons and between nuclei. The force between atoms is given by a sum of all the individual forces, and the fact that the electrons are located outside the atom and the nucleus in the centre.

When two atoms come very close, the force between them is always repulsive, because the electrons stay outside and the nuclei repel each other. Unless both atoms are ions of the same charge (e.g., both negative) the forces between atoms is always attractive at large internuclear distances r. Since the force is repulsive at small r, and attractive at small r, there is a distance at which the force is zero. This is the equilibrium distance at which the atoms prefer to stay.

The interaction energy is the *potential* energy between the atoms. It is negative if the atoms are bound and positive if they can move away from each other. The interaction energy is the integral of the force over the separation distance, so these two quantities are directly related. The interaction energy is a *minimum* at the equilibrium position. This value of the energy is called the *bond energy*, and is the energy needed to separate completely to infinity (the work that needs to be done to overcome the attractive force.) The strongest the bond energy, the hardest is to move the atoms, for instance the hardest it is to melt the solid, or to evaporate its atoms.

Molecular Solids

- Consist of atoms or molecules held together by intermolecular forces (dipole-dipole, dispersion and hydrogen bonds)
  - These forces are weaker than chemical (covalent) bonds. Therefore molecular solids are soft, and have a generally low melting temperature
  - Most substances that are gasses or liquids at room temperature form molecular solids at low temperature (e.g. H_2O, CO_2)

Comparison of bond energies for different types of bonds:

Type of bond	Energy (kJ/mol)
Dispersion (Carbon - carbon van der Waals)	~1.0
Hydrogen bond (strongest dipole-dipole)	~12–16
Ionic	~50–100
Covalent	~100–1000

- The properties of molecular solids also depends upon the *shape* of the molecule

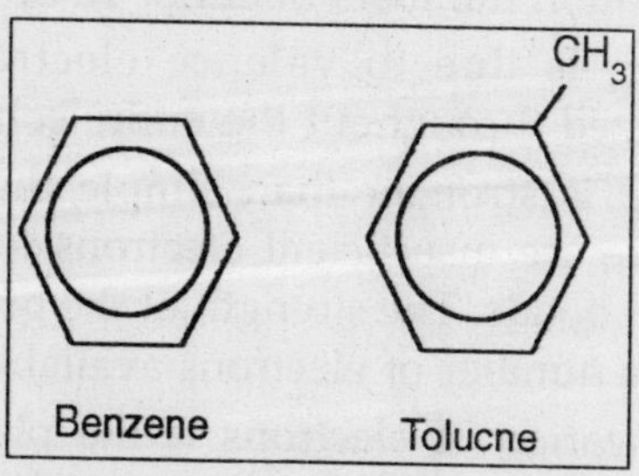

- Benzene (six carbon ring with a symmetrical structure) packs efficiently in three dimensions
- Toluene is related to benzene but has a methyl group attached to one carbon of the ring. It is not

symmetrical and does not pack efficiently. Its melting point is lower than benzene

Covalent Network Solids

- Consist of networks or chains of molecules held together by covalent bonds.
  - Covalent bonds are stronger than intermolecular forces and covalent substances are subsequently harder and have higher melting temperatures.
  - Diamond is a covalent structure of carbon. It is extremely hard and has a melting temperature of 3550C.

Ionic Solids

- Held together by ionic bonds.
- The strength of the ionic interactions depends on the magnitude of the charge of the ions. Thus, NaCl (single charge on both ions) has a melting point of 801C, whereas MgO (2^+, 2^- charge on the ions) has a melting point of 2852C

Metallic Solids

- Consist entirely of metal atoms.
- Typically hexagonal close packed, cubic close packed or body-centered cubic structures. These have coordination numbers of either 12 or 8.
- Bonding is due to valence electrons which are delocalized throughout the entire solid
- Bonding is stronger than simple dispersion forces, but there are insufficient electrons to form ordinary covalent bonds. The strength of the bonding increases with the number of electrons available for bonding
- Delocalization of electrons is the physical basis for the ability of metals to carry electrical current (electrons are free to move about the metal structure)
- The nucleus and inner core of electrons are in a "sea" of delocalized, mobile valence electrons

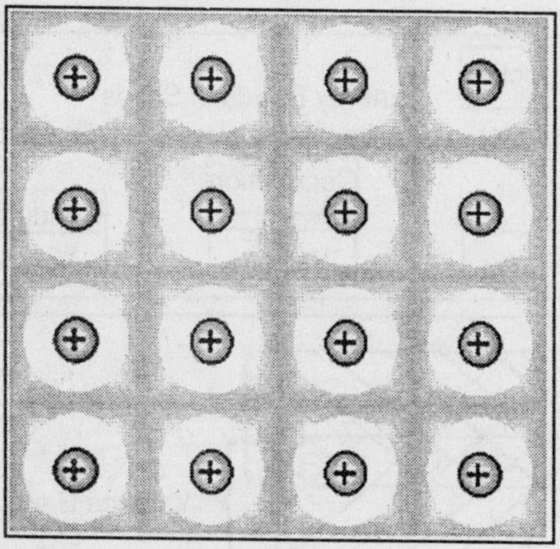

ENERGY BANDS IN SOLIDS

A useful way to visualize the difference between conductors, insulators and semiconductors is to plot the available energies for electrons in the materials.

Instead of having discrete energies as in the case of free atoms, the available energy states form bands. Crucial to the conduction process is whether or not there are electrons in the conduction band.

In insulators the electrons in the valence band are separated by a large gap from the conduction band, in conductors like metals the valence band overlaps the conduction band, and in semiconductors there is a small enough gap between the valence and conduction bands that thermal or other excitations can bridge the gap.

With such a small gap, the presence of a small percentage of a doping material can increase conductivity dramatically.

An important parameter in the band theory is the Fermi level, the top of the available electron energy levels at low temperatures. The position of the Fermi level with the relation to the conduction band is a crucial factor in determining electrical properties.

As shown in Figure below, (a) metals have no forbidden band, because the valence band overlaps the conduction band. Consequently, in metals all electrons will contribute to the comparably high conductivity.

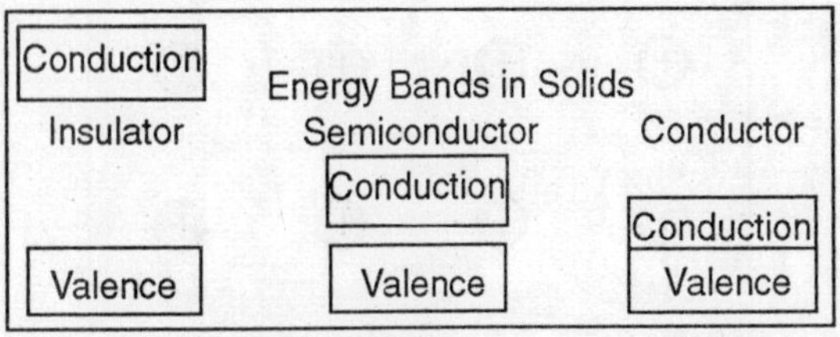

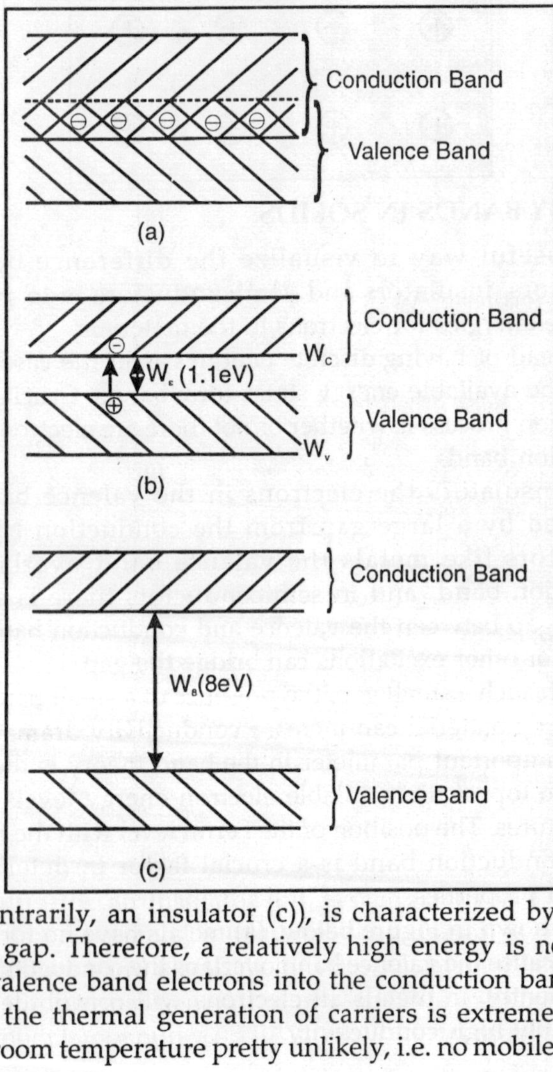

Contrarily, an insulator (c)), is characterized by a large energy gap. Therefore, a relatively high energy is necessary to lift valence band electrons into the conduction band. That is why the thermal generation of carriers is extremely weak and at room temperature pretty unlikely, i.e. no mobile charges are available within the insulator. The semiconductor (b) is neither as a good conductor as the metals are, nor an insulator. The particularity exploited for electronic devices is due to its small intrinsic conductivity, which can be prepared goal-oriented by doping.

To meet the expectations and to illustrate the explanation given above, the energy-band diagram of a p - and n -doped semiconductor under thermal equilibrium will be presented at first (see Figure below).

Obviously, the main differences between these energy-band diagrams concern the relative position of the Fermi-level W_F within the forbidden band, and consequently, the work function W_{WF} of the p -and n -type semiconductor.

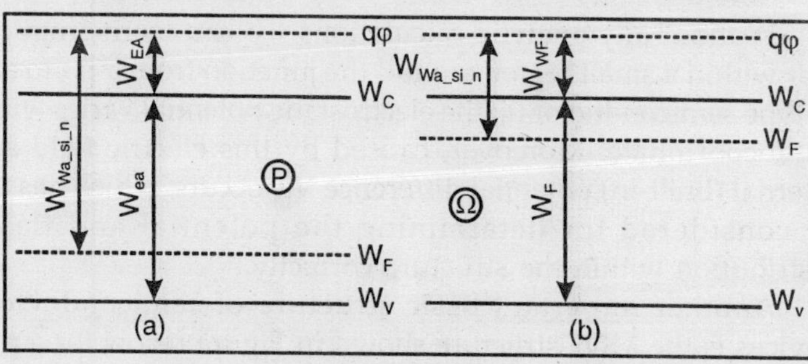

Fig. Energy Band Diagram of a p – and n -doped semiconductor [1], [2]

If we physically implement a p - and n -type semiconductor layer in the same crystal, a so-called pn -junction is formed. To construct its energy-band model we have to bring together the band-diagrams given in Figure above (a) and (b), carefully. In doing so we should consider that W_F is constant throughout the structure. In fact, there are no interface and dipole charges at the stoichiometric junction where the conducting type is technologically changed from p -type to n -type or vice verse. That is why the electrostatic potential within the structure may not exhibit any discontinuities. As mentioned above, the energy equivalent to the electrostatic potential has to stay in parallel to the band edges. Consequently, careful combining the energy-band diagrams of a p - and n -type semiconductor layer results in energy-band diagram of a pn -junction as shown in Figure below.

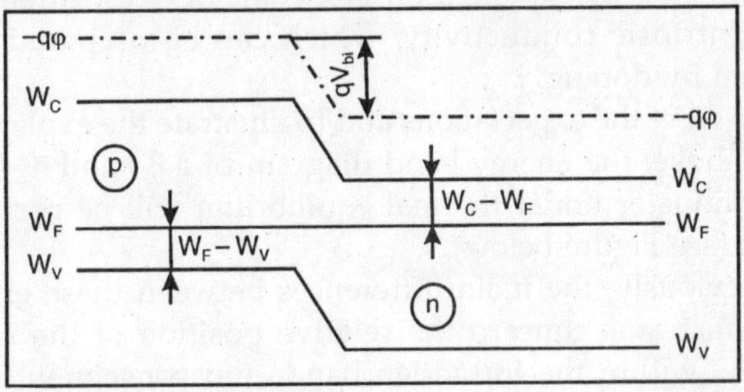

Fig. Energy-Band Diagram of a pn -junction [1], [2]

Without any analytic calculations we can learn from it, that within a small region around the junction from a p - to an n -type semiconductor (ij) the electrostatic potential varies with the I -coordinate. Moreover, caused by this electric field an internal (built-in) potential difference V_{bi} occurs, which has to be considered for determining the potential and field distribution within the structure correctly.

Another important basic structure of semiconductor devices is the MOS-structure shown in Figure below for a p - doped semiconductor.

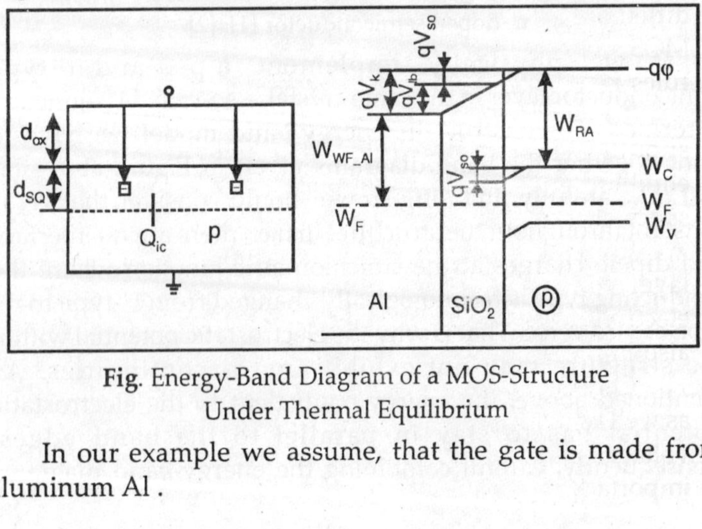

Fig. Energy-Band Diagram of a MOS-Structure Under Thermal Equilibrium

In our example we assume, that the gate is made from aluminum Al .

To come up with the energy band diagram of such a MOS-structure we need the band model for the gate metal (Al) and the p -doped semiconductor. Simply, the metal's energy-band model is given with the Fermi-level W_F and its work function W_{WF}–Al. Separated by a very thin SiO_2 layer a p -doped semiconductor completes the MOS-structure where the semiconductor's band model is well-known already.

Once again, we have to make sure that under thermal equilibrium conditions W_F is constant throughout the complete structure.

Due to the difference in the work function of Si ($W_{WF-Si-p}$) and Al (W_{WF-Al}) we get a linear increase of the potential (potential energy) within the insulator layer. Supposed there are no interface charges located at the interface between the insulator and semiconductor, the electrostatic potential and its slope, may not exhibit any kind of discontinuities at the interface.

Moreover, deep in the semiconductor, i.e. in a sufficient distance from the interface between oxide and semiconductor, the MOS-structure's energy-band diagram has to agree with that of a p -doped semiconductor.

Therefore, we have to accept the band bending as shown in Figure above. Finally, WEA–SiO_2 gives the energy difference to Wc, wherewith the energy-band diagram of the MOS structure can easily be completed, supposed all relevant rules given above are considered carefully.

To sum up what we can learn from this example, we should recognize the following: Already under thermal equilibrium (i.e. without supplied voltages) we can identify a space charge at the interface between SiO_2 and Si.

Roughly speaking, this space charge layer (Q_{Q0} with the layer width d_{Q0}) is due to complex phenomena of contact electronics and leads to a electric field built-in the oxide and also in the semiconductor.

Consequently, a drop of potential across the oxide (V_{io}) as well as the semiconductor (V_{Q0}) will occur. The sum of these voltages is the so-called contact potential v_k, which plays an important role to define the correct boundary conditions for a detailed analysis of the potential and field distribution within the MOS structure with power supply.

CHARGE CARRIERS IN SEMICONDUCTORS

INTRINSIC MATERIAL

Conduction band electrons come from excitation of electrons from the valence band only. The intrinsic carrier density $n_i = p_i$ is a function of temperature and of the band gap energy.

EXTRINSIC MATERIAL

Conduction band electrons and valence band holes come from chemically added dopants: electrons from donors, holes from acceptors. Note that when donors give up an electron to the conduction band, they become positively charged. When acceptors take an electron from the valence band, they become negatively charged. Charged donors and electrons are said to be *ionized,* in analogy with atoms with missing or extra electrons. The charge plays a role in the physics of the device, but since the donors/acceptors cannot move, they do not contribute to the current density.

CONDUCTION OF ELECTRONS/HOLES

An electric field will be produced if we apply a voltage across a piece of semiconductor. Electrons/holes will move in a straight line (on the energy band diagram) if no energy is lost, gaining kinetic energy at the expense of potential energy (analogy: ball rolling down hill). If there are "collisions", then the electron/hole loses energy as it travels.

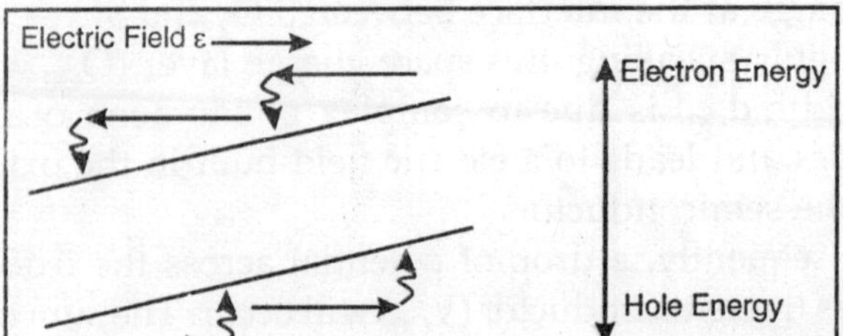

Effective mass: We can account for a lot of detailed physics if we replace the electron/hole mass with the effective mass m_n, m_p.

CARRIER CONCENTRATIONS

We can calculate the equilibrium electron and hole concentrations if we know the Fermi-Dirac distribution function f(E), and the density of states N(E)dE in the conduction and valence bands. The calculation is done with the following approximation: $E - E_F >> kT$, where E is any energy in the conduction or valence band.

$$f(E) = \frac{1}{1 + e^{(E-E_F)}/kT}$$

$$N_C(E) = 4\pi\left(\frac{2m_n}{h^W}\right)^{3/2} (E - E_C)^{1/2} N_V(E) = 4\pi\left(\frac{2m_p}{h^W}\right)^{3/2} (E_V - E)^{1/2}$$

$$n_0 = \int_{Ec}^{\infty} f(E)N_c(E)dE = N_c e - (E_c - E_F)/kT$$

$$p_0 = \int_{-\infty}^{E_V} (1 - f(E))N_V(E)dE = N_V e^{-}(E_F - E_V)/kT$$

$$N_C = 2\left(\frac{2\pi m_n kT}{h^2}\right)^{3/2} \quad N_v = 2\left(\frac{2\pi m_p kT}{h^2}\right)^{3/2}$$

We can also relate the carrier concentrations to the intrinsic Fermi level E_i, as follows.

$$n_0 = n_i e(E_F - E_i)/kT$$

$$p_0 = n_i e^{-}(E_i - E_F)/kT$$

Finally, we have a very important constraint on the carrier concentrations.

$$n_0 p_0 = n^2{}_i = N_C N_V e^{-E_g}/kT$$

Ideas

N_C and N_V can be calculated from the equations given, but at room temperature, we can more easily look them up in the table given out at the beginning of the semester.

For an n-type semiconductor, $n_o > p_o$, and the Fermi level is closer to the conduction band edge than it is to the valence band edge.

For a p-type semiconductor, $p_o > n_o$, and the Fermi level is closer to the valence band edge than it is to the conduction band edge.

In *any* semiconductor *at equilibrium*, regardless of doping density, the product of the electron and hole concentrations is a constant: $n_o p_o = n_i^2$.

If the doping densities are known (as is usually the case) we can find the carrier concentrations in a semiconductor as follows.

For n-type semiconductors ($N_d > N_a$):

$$n_0 = \frac{N_d - N_a}{2} + \sqrt{\frac{N_d - N_a}{2} + n^2{}_i} \; p_0 = n_i^2 / n_0$$

For p-type semiconductors ($N_a > N_d$):

$$p_0 = \frac{N_a - N_d}{2} + \sqrt{\frac{N_d - N_d}{2} + n^2{}_i} \; n_0 = n_i^2 / p_0$$

Note that in semiconductors with band gaps around 1 eV or larger, the intrinsic carrier density at room temperature is *much* smaller than typical doping densities, so that we can write the following equations.

This is true for Si at room temperature, for example:

For n-type:

$$n_0 = N_d - N_a p_0 = n^2{}_i / n_0$$

For p-type:

$$p_0 = N_a - N_d n_0 = n^2{}_i / p_0$$

DRIFT CURRENT

Electrons and holes have a thermal velocity $v_{thermal}$, due to random thermal energy. When an electric field is applied, they move with an average velocity v_{drift}. The drift velocity, drift current density J, mobility m, and conductivity s are related as follows. Expressions are given for electrons, but similar expressions hold for holes.

$$\frac{1}{2} m^* v_{thermal}^2 = \frac{3}{2} kT$$

$$< v >= v_{drift} = -\mu_n \varepsilon$$

$$J_n^{DRIFT} = \sigma_n \varepsilon = q \mu_n n \varepsilon$$

$$J_n^{DRIFT} = J_n^{DRIFT} + J_p^{DRIFT} = \sigma\varepsilon \qquad \sigma = qu_n n + q\mu_p p$$

The conductivity σ, resistivity ρ, and resistance R are related as follows.

$$\sigma = 1/\rho R = \rho\frac{L}{A}$$

The resistance R is from the more familiar form of Ohm's law: v = iR.

Ideas

The mobility is a function of temperature and defect density, where "defect" includes intentionally added dopants. We can look up mobility on a chart such as that given in Fig. 3–23. At temperatures other than room temperature, we must have a chart or otherwise be given the mobility.

Electrons and holes scatter from both ionized donors and ionized acceptors. Therefore, the presence of donors and acceptors will affect the mobility of both holes and electrons. In determining the mobility from Fig shown for example, we must look up the mobility at the value $N_d + N_a$, i.e., the sum of the doping densities.

INVARIANCE OF THE FERMI LEVEL

At equilibrium, the Fermi level is constant throughout a given piece of material. This idea will be important in drawing energy band diagrams for devices comprised of more than one material.

DRIFT OF CARRIERS IN ELECTRIC AND MAGNETIC FIELDS

The motion of a carrier drifting in a semiconductor due to an applied electric field, e, is illustrated in Figure below. The field causes the carrier to move with a velocity, *v*.

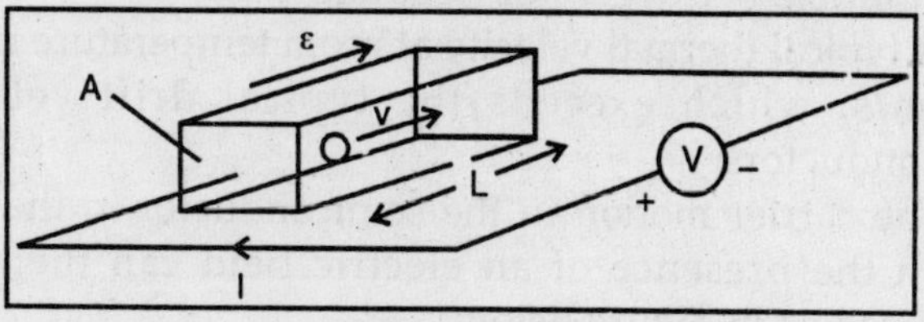

Assuming that all the carriers in the semiconductor move with the same velocity, the current can be expressed as the total charge in the semiconductor divided by the time needed to travel from one electrode to the other, or:

$$I = \frac{Q}{t_r} = \frac{Q}{L / v}$$

where t_r is the transit time of a particle, traveling with velocity, v, over the distance L.

The current density, J, can then be rewritten as a function of the charge density, r :

$$\vec{J} = \frac{\vec{I}}{A} = \frac{Q}{AL}\vec{v} = \rho\vec{v}$$

. If the carriers are negatively charged electrons, the current density equals:

$$\vec{J} = -qn\vec{v}$$

while for positively charged holes it is:

$$\vec{J} = qp\vec{v}$$

Where n and p are the electron and hole density in the semiconductor.

It should be noted that carriers do not follow a straight path along the electric field lines. Instead they bounce around in the semiconductor and constantly change direction and velocity due to scattering.

This behaviour occurs even when no electric field is applied and is due to the thermal energy of the carriers. Thermodynamics teaches us that electrons in a non-degenerate and non-relativistic electron gas have a thermal energy of $kT/2$ per particle per degree of freedom.

A typical thermal velocity at room temperature is around 10^7 cm/s, which exceeds the typical drift velocity in semiconductors.

The carrier motion in the semiconductor in the absence and in the presence of an electric field can therefore be visualized as in Figure below.

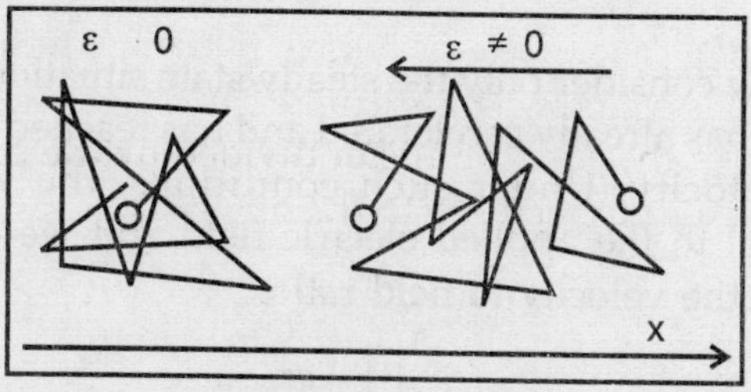

Fig. Random Motion of Carriers in a Semiconductor with and without an Applied Electric Field.

In the absence of an applied electric field, the carrier exhibits random motion and the carriers move quickly through the semiconductor and frequently change direction.

When an electric field is applied, the random motion still occurs but in addition, there is on average a net motion along the direction of the field. Due to their different electronic charge, holes move on average in the direction of the applied field, while electrons move in the opposite direction.

We now analyse the carrier motion considering only the average velocity, $\langle \vec{v} \rangle$ of the carriers. Applying Newton's law, we state that the acceleration of the carriers is proportional to the applied force:

$$\vec{F} = m\vec{a} = m\frac{\langle \vec{v} \rangle}{dt}$$

The force consists of the difference between the electrostatic force and the scattering force due to the loss of momentum at the time of scattering. This scattering force equals the momentum divided by the average time, t_c, between scattering events or collisions, so that:

$$\vec{F} = q\vec{\varepsilon} - \frac{m\langle \vec{v} \rangle}{\tau_c}$$

where a particle with charge q was assumed.

Combining both relations yields an expression for the average particle velocity:

$$q\vec{\varepsilon} = m\frac{d\langle \vec{v} \rangle}{dt} + \frac{m\langle \vec{v} \rangle}{\tau_c}$$

We now consider only the steady state situation in which the particle has already accelerated and has reached a constant average velocity. Under such conditions, the velocity is proportional to the applied electric field and we define the mobility as the velocity to field ratio:

$$\mu \overset{\Delta}{=} \frac{|\bar{v}|}{|\bar{\varepsilon}|} = \frac{q\tau_c}{m}$$

The mobility of a particle in a semiconductor is therefore expected to be large if its mass is small and the time between scattering events is large.

The drift current, described as above, can then be rewritten as a function of the mobility, yielding:

$$\bar{J}_n = qn\mu_n\bar{\varepsilon}$$

For electrons and,

$$\bar{J}_p = qp\mu_p\bar{\varepsilon}$$

For holes,

Throughout this derivation, we simply considered the mass, *m*, of the particle. However in order to incorporate the effect of the periodic potential of the atoms in the semiconductor we must use the effective mass, m^*, rather than the free particle mass:

$$\mu = \frac{q\tau_c}{m^*}$$

Example: Electrons in undoped gallium arsenide have a mobility of 8,800 cm^2/V-s. Calculate the average time between collisions. Calculate the distance traveled between two collisions (also called the mean free path). Use an average velocity of 10^7 cm/s.

Solution: The collision time, t_c, is obtained from:

$$\tau_c = \frac{\mu_n m_e^*}{q} = \frac{0.88 \times 0.067 \times 9.1 \times 10^{-31}}{1.6 \times 10^{-19}} = 0.34 ps$$

where the mobility was first converted in MKS units.

The mean free path, *l*, equals:

$$l = v_{average}\tau_c = 10^7 \times 0.34 \times 10^{-12} = 34 nm$$

Chapter 2

Characteristics of Diode

HIGH VACUUM DIODE

A vacuum diode is an electron valve that supports current flow through it in only one direction. It is composed of two electrical "parts" or elements and the evacuated glass envelope. Additionally, there is usually a heater, a coil of special alloy wire that can be driven to red heat by a modest application of voltage. (A vacuum diode without a heater would be a cold cathode diode.) The heater is coiled inside a cylindrical element called the cathode.

The cathode is composed of a special alloy, or, more properly is coated with an alloy, that has high thermal emission characteristics. That means that if it's heated sufficiently, electrons actually leave the metallic crystal structure of the metal (boil off) and hang around outside in the space nearby forming what is called a space charge. When we built the tube, we pumped all the air out, sealed the glass, and then fired a "getter" inside the tube to chemically bind the remaining bit of gas that was there.

The tube really is highly evacuated. That's important because we don't want all those light weight electrons bumping into gas atoms and molecules when they are hanging around in the space charge, and especially when transiting the gap between the cathode and the other element inside, the anode or plate.

If a voltage is applied across the cathode and plate, and the plate is negative, no current will flow in the tube. Electrons flow from negative to positive, but, darn it, they can't leave the plate to go to the cathode. That's called reverse biased. It's hooked up "backwards" like that. But if the cathode is connected to negative and the positive connected to the plate, electrons that were in the space charge will be driven away from the cathode and attracted to the plate. Current will flow in this direction. That's forward bias.

The hot cathode emits electrons in droves. Many more than are needed. They build up in the volume around the cathode and form a large pool of electrons just waiting for something positive to happen.

This pool of electrons is called the space charge. When the plate (anode) is made positive some of the electrons are attracted to it. They impact on it, are absorbed into the metal and electrons flow out of the plate connection into the battery or what ever provided the positive voltage. The other end of the voltage source must be connected back to the cathode in some way.

If the polarity is reversed which makes the plate negative with respect to the cathode the electrons in the space charge are repelled away from the plate and no conduction takes place. The plate is cold and is made of a metal that is a very poor emitter of electrons.

The voltages are not nearly as high as those used by Crooks so there is no cold emission from the plate. This makes the diode conduct current in only one direction.

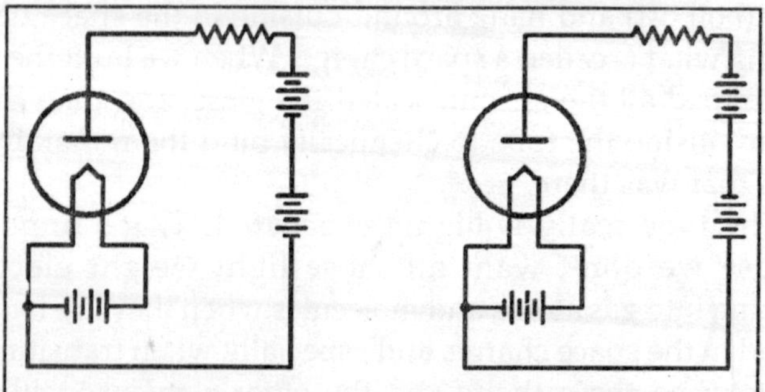

Heavy Conduction. No Conduction.

The cathode of a tube is coated with a very special material that emits electrons very easily. It can readily be damaged by operating above or below the proper temperature or by being "poisoned" by the residual air in the tube.

The space charge serves another vital function. There is no such thing as a perfect vacuum. In fact the space in near earth orbit is a better vacuum than can be made on the surface of earth. Farther out in space the vacuum is even better. What this means is there are lots of air molecules running around inside of a vacuum tube.

The space charge protects the cathode from these air molecules. Some of these molecules lose an electron and become positive ions. They are attracted to the negative cathode and if it weren't for the space charge they would impact on the surface of the cathode doing considerable damage. Over time the ability of the cathode to emit electrons would be seriously impaired rendering the tube useless. As these ions run through the space charge they gain electrons from the many collisions and become negative ions. Their direction is reversed turning them away from the delicate cathode.

INDIRECTLY HEATED CATHODES

Most vacuum tubes have a basically cylindrical configuration. Figure two shows how a filament (directly heated) cathode type diode is made along with a heater type (indirectly heated) cathode type diode.

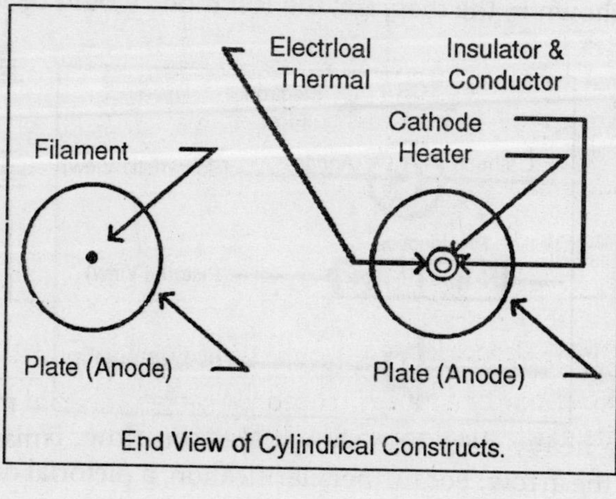

End View of Cylindrical Constructs.

SEMICONDUCTOR DIODE

INTRODUCTION

If we join a section of N-type semiconductor material with a similar section of P-type semiconductor material, we obtain a device known as a PN JUNCTION. (The area where the N and P regions meet is appropriately called the junction.) The usual characteristics of this device make it extremely useful in electronics as a diode rectifier.

The diode rectifier or PN junction diode performs the same function as its counterpart in electron tubes but in a different way.

The diode is nothing more than a two-element semiconductor device that makes use of the rectifying properties of a PN junction to convert alternating current into direct current by permitting current flow in only one direction.

The schematic symbol of a PN junction diode is shown in figure below. The vertical bar represents thecathode (N-type material) since it is the source of electrons and the arrow represents the anode. (P-type material) since it is the destination of the electrons.

The label "CR1" is an alphanumerical code used to identify the diode. In this figure, we have only one diode so it is labeled CR1 (crystal rectifier number one). If there were four diodes shown in the diagram, the last diode would be labeled CR4.

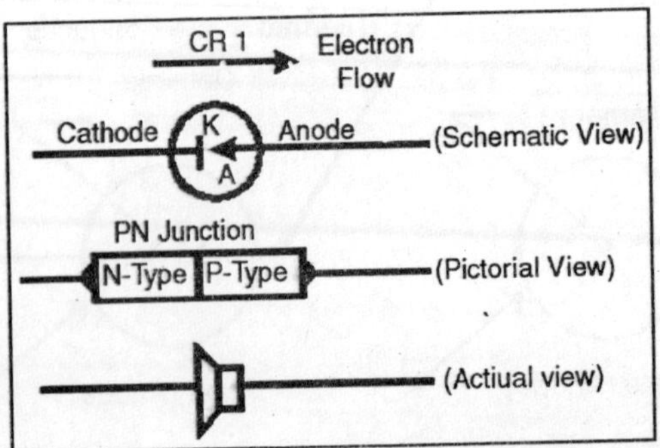

The heavy dark line shows electron flow. Notice it is against the arrow. For further clarification, a pictorial diagram of a PN junction and an actual semiconductor (one of many types) are also illustrated.

CONSTRUCTION

Merely pressing together a section of P material and a section of N material, however, is not sufficient to produce a rectifying junction.

The semiconductor should be in one piece to form a proper PN junction, but divided into a P-type impurity region and an N-type impurity region.

This can be done in various ways. One way is to mix P-type and N-type impurities into a single crystal during the manufacturing process.

By so doing, a P-region is grown over part of a semiconductor's length and N-region is grown over the other part. This is called a GROWN junction and is illustrated in view A of figure below.

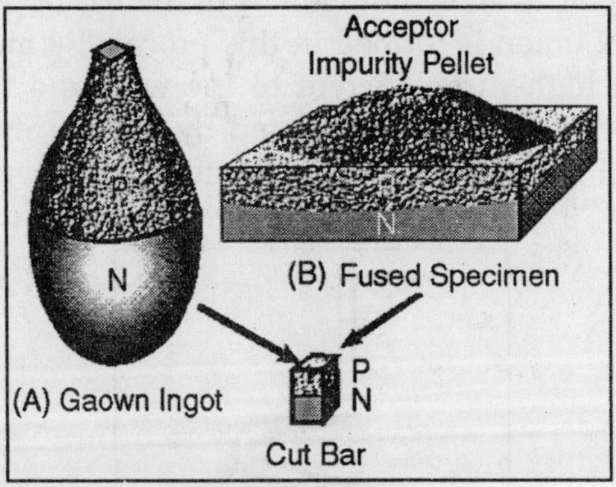

Another way to produce a PN junction is to melt one type of impurity into a semiconductor of the opposite type impurity. For example, a pellet of acceptor impurity is placed on a wafer of N-type germanium and heated.

Under controlled temperature conditions, the acceptor impurity fuses into the wafer to form a P-region within it, as shown in view B of figure above. This type of junction is known as an ALLOY or FUSED-ALLOY junction, and is one of the most commonly used junctions. In figure shown below, a POINT-CONTACT type of construction is shown.

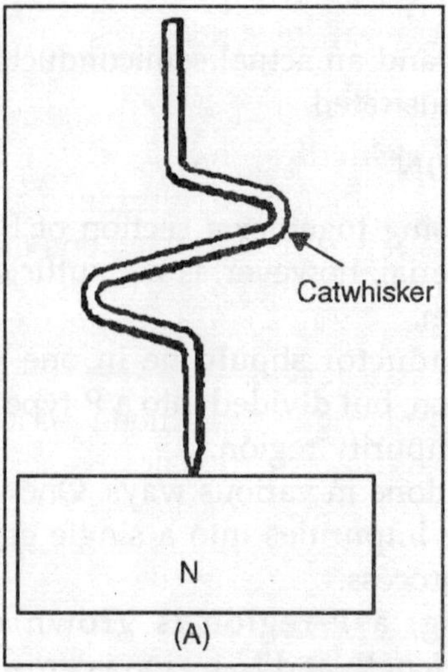

It consists of a fine metal wire, called a cat whisker, that makes contact with a small area on the surface of an N-type semiconductor as shown in view A of the figure.

The PN union is formed in this process by momentarily applying a high-surge current to the wire and the N-type semiconductor. The heat generated by this current converts the material nearest the point of contact to a P-type material (view B).

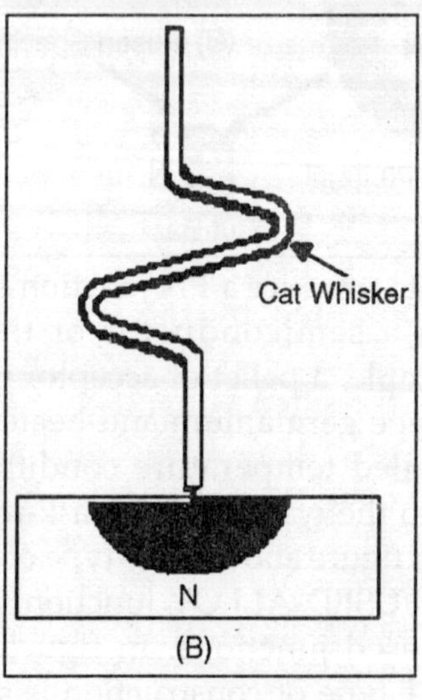

TYPES OF DIODES

The first diode in figure below is a semiconductor diode which could be a small signal diode of the 1N914 type commonly used in switching applications, a rectifying diode of the 1N4004 (400V 1A) type or even one of the high power, high current stud mounting types. You will notice the straight bar end has the letter "k", this denotes the "cathode" while the "a" denotes anode.

Current can only flow from anode to cathode and not in the reverse direction, hence the "arrow" appearance. This is one very important property of diodes.

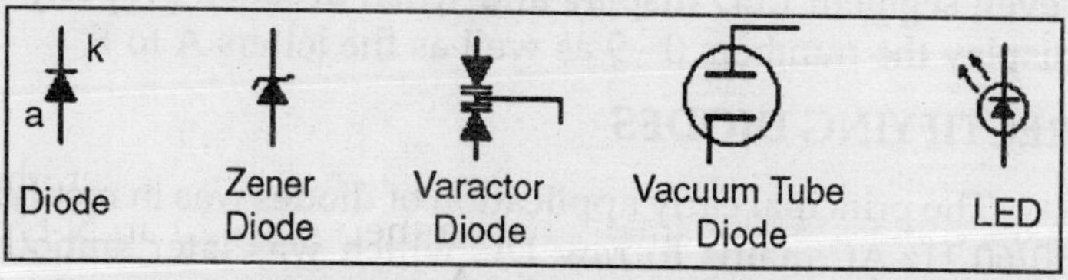

The second of the diodes is a zener diode which are fairly popular for the voltage regulation of low current power supplies. Whilst it is possible to obtain high current zener diodes, most regulation today is done electronically with the use of dedicated integrated circuits and pass transistors.

The next of the diodes in the schematic is a varactor or tuning diode. Depicted here is actually two varactor diodes mounted back to back with the DC control voltage applied at the common junction of the cathodes.

These cathodes have the double bar appearance of capacitors to indicate a varactor diode. When a DC control voltage is applied to the common junction of the cathodes, the capacitance exhibited by the diodes (all diodes and transistors exhibit some degree of capacitance) will vary in accordance with the applied voltage.

A typical example of a varactor diode would be the Philips BB204G tuning diodes of which there are two enscapsulated in a TO-92 transistor package.

At a reverse voltage Vr (cathode to anode) of 20V each diode has a capacitance of about 16 pF and at Vr of 3V this capacitance has altered to about 36 pF. Being low cost diodes, tuning diodes have virtually replaced air variable capacitors in radio applications today.

The next diode is the simplest form of vacuum tube or valve. It simply has the old cathode and anode. These terms were passed on to modern solid state devices. Vacuum tube diodes are mainly only of interest to restorers and tube enthusiasts.

The last diode depicted is of course a light emitting diode or LED. A led actually doesn't emit as much light as it first appears, a single LED has a plastic lens installed over it and this concentrates the amount of light.

Seven LED's can be arranged in a bar fashion called a seven segment LED display and when decoded properly can display the numbers 0 - 9 as well as the letters A to F.

RECTIFYING DIODES

The principal early application of diodes was in rectifying 50/60 Hz AC mains to raw DC which was later smoothed by choke transformers and/or capacitors.

This procedure is still carried out today and a number of rectifying schemes for diodes have evolved, half wave, full wave and bridge rectifiers.

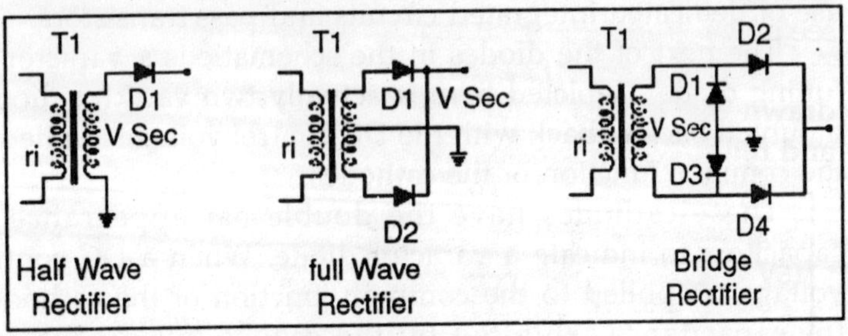

Fig. Rectifying Diodes

As examples in these applications the half wave rectifier passes only the positive half of successive cycles to the output filter through D1. During the negative part of the cycle D1 does not conduct and no current flows to the load.

In the full wave application it essentially is two half wave rectifiers combined and because the transformer secondary is centre tapped, D1 conducts on the positive half of the cycle while D2 conducts on the negative part of the cycle. Both add together. This is more efficient.

The full wave bridge rectifier operates essentially the same as the full wave rectifier but does not require a cetre tapped transformer. Further discussion may be seen on the topic power supplies.

A further application of rectifying diodes is in the conversion or detection of rf modulated signals to audio frequencies. Typical examples are am modulated signals being detected and early detection schemes for fm also used diodes for detecting modulation.

VOLTAGE REGULATING DIODES

For relatively light current loads zener diodes are a cheap solution to voltage regulation. Zener diodes work on the principle of essentially a constant voltage drop at a predetermined voltage (determined during manufacture).

An example is a Philips BZX79C12 type with a regulation range between 11.4V and 12.7V but typically 12V and a total power dissipation of 500 mW in a DO-35 package. The dissipation can be extended by using a series pass transistor, see power supplies.

Notice in figure 3 there is a resistor to miminmise current drawn but mainly as an aid to dropping the supply voltage and reducing the burden on the zener diodes.

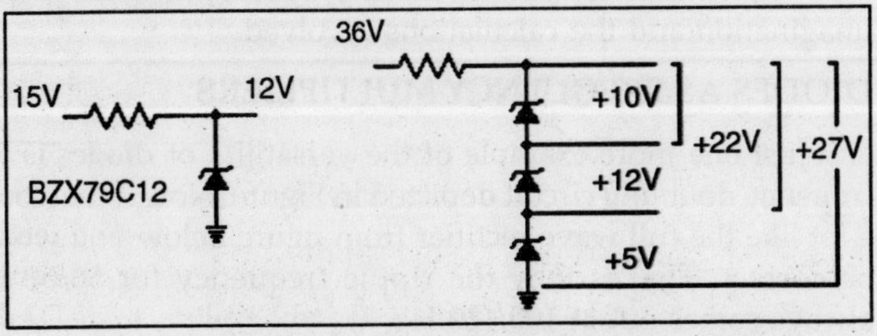

Fig. Zener Voltage Regulation Diodes

In the second schematic of figure above we have three zener diodes in series providing voltages of 5V, 10V, 12V, 22V and 27V all from a 36V supply. This configuration is not necessarily recommended especially when the current being drawn is seriously mismatched between voltages. It is presented purely out of interest.

VARACTOR OR TUNING DIODES

These types of diodes work on the principle that all diodes exhibit some capacitance. Indeed the zener diode BZX79C12 quoted above has, according to the data book, a capacitance of 65-85 pF at 0V and measured at 1 Mhz.

For AM Radio band applications a specific diode has been devised. The Philips BB212 in a TO-92 case is one such type. Each of the diodes has a capacitance of 500 - 620 pF at a reverse bias of 0.5V and <22 pF at 8V.

This diode's capacitance ratio is quoted at 22.5:1 which could not be achieved easily if at all with an air variable capacitor.

Several obvious advantages come immediately to mind, a small transistor type package, very low cost, ease of construction on a circuit board, can be mounted away from heat generating devices, frequency determining circuitry entirely dependent upon resistor values and ratios, DC voltage control can be either from frequency synthesiser circuits or perhaps a multi-turn potentiometer.

Such a potentiometer aids band spreading and fine tuning if two potentiometers are used. The only real limitation is your imagination and the calculations involved.

DIODES AS FREQUENCY MULTIPLIERS

Just one more example of the versatility of diodes is the frquency doubling circuit depicted in Figure. Now if that looks a lot like the full wave rectifier from figure below you would be correct. That is why the ripple frequency for 50/60 Hz always comes out at 100/120 Hz.

Here the input is a wide band transformer and the signal passes to a full wave rectifier comprising two 1N914 diodes. The DC component caused by the rectification passes to ground through RFC which of course presents a high impedance to the rf porion of the signal but essentially a short circuit for DC.

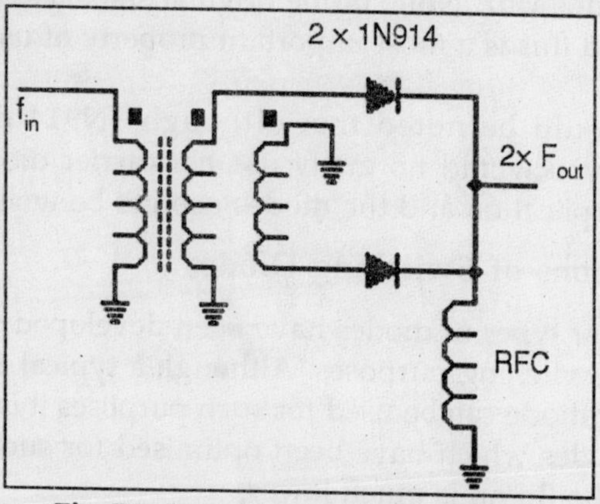

Fig. Diodes as Frequency Multipliers

The original signal should be down about 40 dB and with this type of circuit there would be a loss of somewhere around 7.5 dB so the 2 X signal would require further amplification to restore that loss.

DIODES AS MIXERS

With some subtle re-arrangement to figure 4 we can get the circuit to function as a two diode frequency mixer. Note that there are other diode arrangements as well in this application.

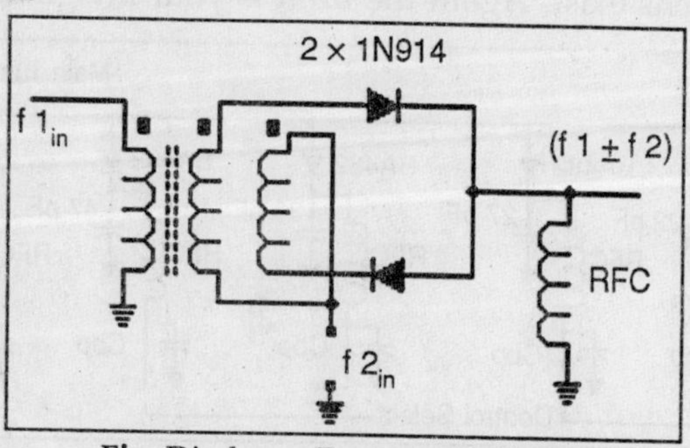

Fig. Diodes as Frequency Mixers

The diodes here act as switches and it can be mathematically shown that only the sum and difference signals will result. For example, if F1 was 5 Mhz and F2 was 3 Mhz then the sum and difference signals from the diodes would be 8 Mhz and 2 Mhz. None of the original signals appear at the output and this is a most important property of using diodes as mixers.

It should be noted that although 1N914 diodes are depicted you would normally use hot carrier diodes in any serious application and the diodes need to be well matched.

Applications of Switching Diodes

Similar types of diodes have been developed specifically for band switching purposes. Although a typical 1N914 type switching diode can be used for such purposes it is preferable to use diodes which have been optimised for such purposes because the Rd on is much lower.

This means the diode resistance Rd can have a serious affect on rf circuits in particular the "Q" of a tuned circuit. One example of a low Rd device is the Philips BA482 diode used for band switching in television tuners.

It has a typical Rd of 0.4 ohms at a forward current of 10 mA. In figureabove we have one application where switching diodes operate.

All diodes serve to switch in or out capacitors in the diagram which is presented here just to illustrate one single application of switching diodes, many, many more applications exist. Again the limit is your imagination.

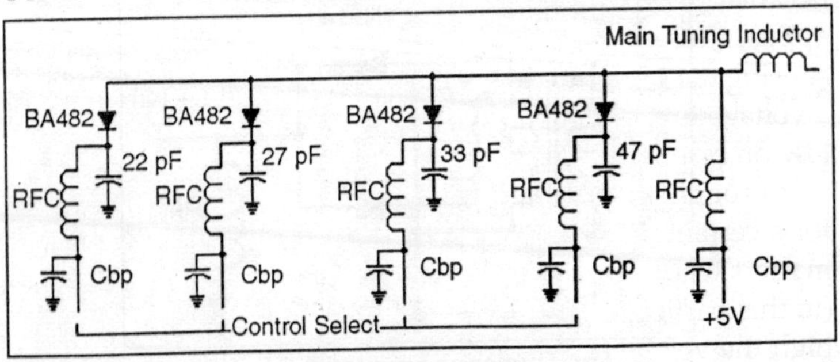

Fig. Applications of Switching Diodes

The switching diodes in figure switch in or out successively higher values of capacitors as each control select line is "grounded". The voltage from the +5V feed line proceeds through the diode at DC thus opening the diode and making it appear "transparent" for rf purposes.

The capacitor with the value attached is then "switched" into circuit. Other components marked RFC and Cbp are chokes and bypass capacitors for "clean" switching. The bypass capacitors and choke values would be determined by the frequency of operation.

We could just have easily have switched inductors instead of capacitors. Note why Rd is quite important on overall circuit performance.

If we were using inductors the diode resistance Rd would have a significant affect on inductor "Q" which in turn would affect filter performance, if it was in fact an LC filter application.

SWITCHING DIODES IN LOGIC CIRCUITS

If you you completed the tutorial on digital basics you should be aware of binary numbers. There are a whole range of digital building blocks available and just by way of one illustration of using diodes we have presented the 74HC4040 twelve stage binary ripple counter (there are others with varying number of stages).

In the schematic of figure we have this counter which divides by successive division of two for twelve stages. Initially because there is no voltage drop across the resistor a high appears on all anodes as well as on pin 4 the master reset causing the counter to reset forcing all outputs low and in turn a voltage drop across each diode and across the resistor and a low on reset.

Progressively each of the outputs change from low to high for a certain period of time and without unduly complicating matters when all outputs as selected by our diode combination (in this particular case 1 + 2 + 32 + 64 = 99) are simultaneously high the voltage drop across the resistor will cease and cause pin 11 (reset which was formerly low) to go high and reset all the internal ripple counters.

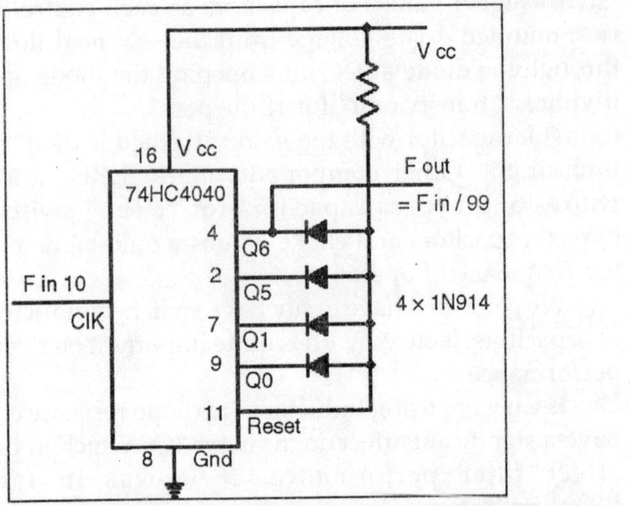

Fig. Applications of Switching Diodes in Digital Logic Circuits

At the same time pin 4 changes state also with reset. It can been shown this happens once every 99 periods. Simply by placing diodes on the right outputs we can select to divide by any number up to 4095 using this particular counter.

LIGHT-EMITTING-DIODES OR LED'S

Many circuits use a led as a visual indicator of some sort even if only as an indicator of power supply being turned on. A sample calculation of the dropping resistor is included in figure

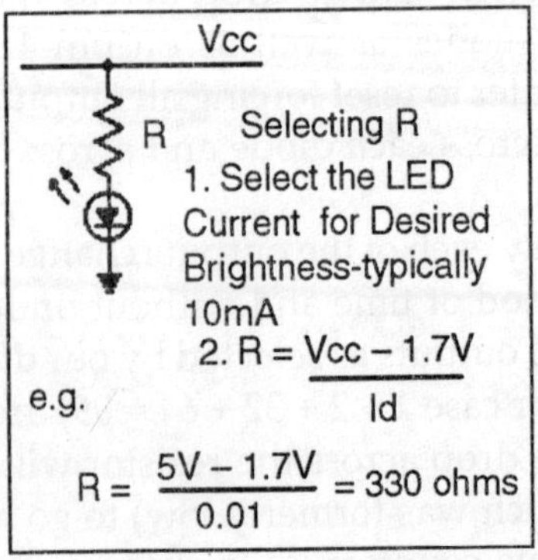

Fig. Connecting Light Emitting Diodes (LED's) to Supply

Most leds operate at 1.7V although this is not always the case and it is wise to check. The dropping resistor is simply the net of supply voltage minus the 1.7V led voltage then divided by the led brightness current expressed as "amps" (ohms law). Note the orientation of both cathode and anode with respect to the ground end and the supply end. Usually with a led the longer lead is the anode.

IDEAL DIODE

DIODES

Diodes are different and useful electrical components. Diodes are used in many applications like the following.

- Converting AC power from the 60Hz line into DC power for radios, televisions, telephone answering machines, computers, and many other electronic devices.
- Converting radio frequency signals into audible signals in radios.

Here are the goals for this lesson unit:

- Given a circuit with a diode,
  - Be able to use a simplified model of a diode to predict when current flows through the diode, and when it does not.
  - Be able to use information about current flowing to predict other behaviour in a circuit.

IDEAL DIODE MODEL

The perfect diode model must have the following characteristics:

- It is a two-port device (at least external to the model/ subcircuit for the device).
- When the externally applied voltage from Cathode to Anode terminals is negative, essentially no internal current must flow (subject of course to numerical error limit considerations).
- When condition 2. is not met, the device current (and voltage) must be determined by the external circuitry, the device behaving as a short circuit (subject of

course to the same numerical error limit considerations).

Using an internal opposing voltage source, one can construct a model and a test circuit for the first pass at an ideal diode as shown in Figure following:

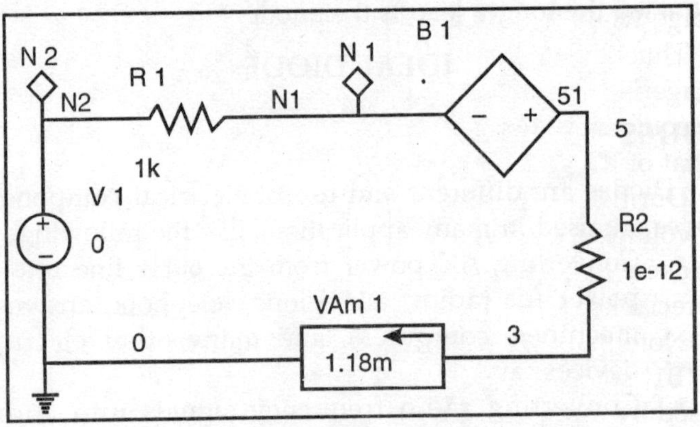

Fig. Ideal Diode Test Circuit

Performing a transient analysis on the circuit produces the results in Figure. The results are as expected, with the simulated ideal diode consisting of the 'B1' and 'R2' elements, having its anode at node 'N1." and its cathode at node '3'.

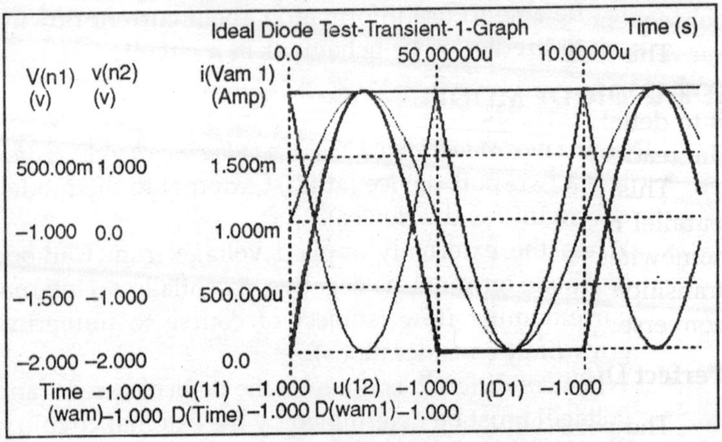

Fig. Ideal Diode Test Circuit

The operation of the diode circuit is straightforward. The applied voltage is opposed by the 'B1' voltage generator. The voltage of the 'B1' generator is the resultant of two levels. The first part is the absolute value of the voltage from node 'N1' to ground. The second is the uramp value of the voltage from node 'N1' to ground. The uramp value is the value of the 'N1' voltage when it is greater than zero.

Thus, during the half-sinusoid interval when the applied voltage is (or would be) positive, the voltage is unopposed. The equivalent diode 'conducts' with an 'ON' resistance equal to that of R2.

During the second, negative half of the rectified sinusoid, the voltage generator 'B1' causes the voltage at 'N1' to be opposed by an essentially equal level, resulting in no appreciable current flow. This is of course an open circuit.

Note that the currents indicated by the ammeter and by the 'B1' source value are 180 degrees out of phase. This is caused by the sense of the 'B1' generator in the circuit. For a voltage source, within that source, 'positive' conventional current flows from negative to positive levels.

So far so good, however the sense of the 'B1' generator is negative when it is (for all essential purposes) non-zero. This causes the 'B1' current to seem to be wrong. However as this will be external to the diode model/subcircuit it is of no consequence.

This ideal diode could be added to the library as a device, perhaps as a symbol of a normal diode with a letter 'I' beside it to denote it as an ideal diode. This is left as an exercise for the reader.

This device can be associated with series resistance, parallel resistance and or capacitance as needed to make it somewhat less ideal and ease simulation difficulties at the transition points from conduction to non-conduction and the converse.

Perfect Diode Model

The perfect diode model would seem to be very similar to that of the ideal diode. The differences are that it should exhibit a forward voltage drop VF when it conducts with an essentially zero incremental resistance state, and when reverse biased with a sufficiently large voltage, it will conduct at a voltage VR with an essentially zero incremental resistance state. At first glance, it would appear that one could use one ideal diode with a fixed voltage source of VF in series opposition to the forward voltage, together with an anti-parallel ideal diode with a fixed voltage source of VR in series with it in its 'forward' direction.

The first diode would produce the 'ON' voltage drop of VF when its net applied voltage, Vin-VF was greater than zero. In the reverse direction, the anti-parallel diode would conduct when the overall diode voltage was less than VR. But, it does not work.

The forward characteristics can be simulated by adding a fixed voltage in series with an ideal diode, as can the reverse characteristics by a single anti-parallel diode with an associated VR. But combining the two will not work. A schematic of this failed attempt to create a (partial) perfect diode in that manner is shown in Figure.

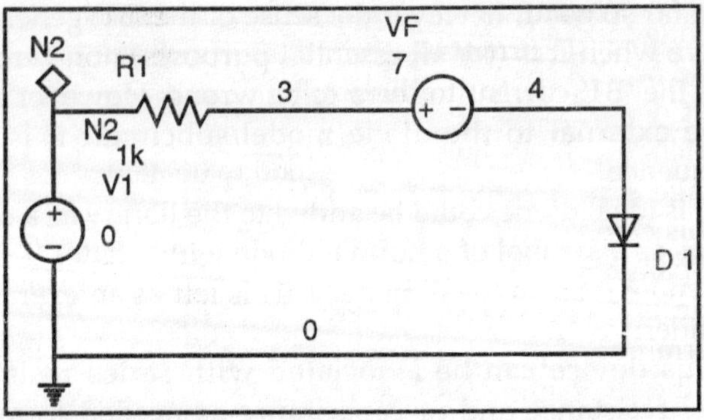

Fig. Perfect Diode fwd Conduction Element

Device D1 is a subcircuit representation of the ideal diode. VF is the forward drop of the simulated perfect diode, set at 0.7V. This illustrates the forward voltage conduction portion of the perfect diode and it works as expected. Figure below shows the graph of this circuit, with the forward current through the device as the current through voltage source VF. It can be verified that the breakdown characteristics can be modeled by a reversed D1 and VF (becoming VR). The problem comes when one tries to parallel these items. The circuit, as shown in Figure, which will not work properly.

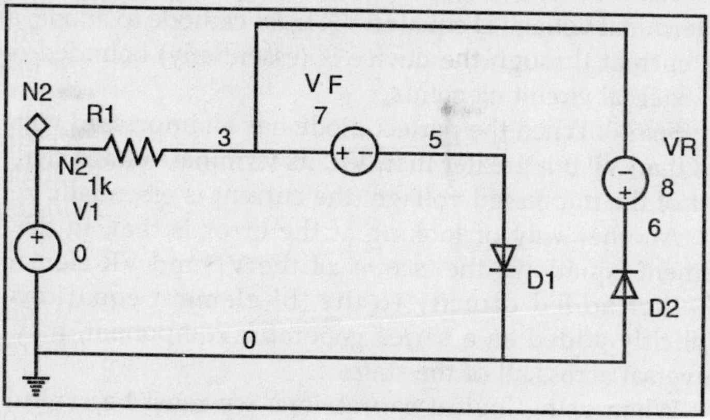

Fig. (FAILED) Perfect Diode Model

The circuit will not converge. Why not? There are TWO reasons for this circuit failure. The first is due to the mixing of essentially ideal branches 3-5-0 and 3-6-0. The second is due to the improper modeling of the perfect diode itself.

Considering the first error, it must be remembered that IF a circuit element passes no current, either it has no net voltage impressed, or it has an essentially infinite impedance. In this case, when diode D1 is NOT forward biased, and the anti-parallel diode is NOT forward biased (the overall perfect diode is not conducting), the voltage at node 3 cannot be determined.

It cannot simultaneously be both VIN - VF, and VIN +VR or any combination thereof, as D1 and D2 are essentially open circuited. Considering the second error, the model we assumed without thought, was wrong. There are THREE states of interest for the perfect diode. The ideal diode has only two states. The three states for the perfect diode are as follows:

State 1: When the perfect diode has an impressed voltage greater than VF, from cathode to anode, it enters a state wherein its terminal voltage is equal to VF from cathode to anode, and the current through the device is (essentially) bounded only by external circuit elements

State 2: When the perfect diode has an impressed voltage less than VR, from cathode to anode, it enters a state wherein its terminal voltage is equal to VR from cathode to anode, and the current through the device is (essentially) bounded only by external circuit elements

State 3: When the perfect diode has an impressed voltage less than VF but greater than VR, its terminal voltage shall be that of the impressed voltage (the current is essentially zero)

Another way of looking at the error is that, in the 'B' element equations, the 'scope' of the VF and VR elements, whether added directly to the 'B' element equations or explicitly added as a series generator component, may be universal across all of the states.

When using logical expressions we must be careful to ensure that all of the device states are limited properly in scope. Logically, we need to implement a voltage generator 'B' element expression which has the following effect:

Regions	Condition	Device Voltage
I	VIN > VF	Vd = VF
II	not(VIN > VF) and not(VIN < VR)	Vd = VIN
III	VIN < VR	Vd = VR

There was a specific reason for writing the region II conditions as shown. Remember that the 'u' function returns a unity value when its argument is GREATER than zero.

As mentioned in a previous article, the logical inverse of this condition is a function with a value of unity when its argument is EQUAL TO or LESS THAN zero.

Practically speaking, in many cases one could merely invert the original argument to arrive at this, however it is easier to visualize and always correct to create a logical inverse function.

Recalling what was previously discussed, if a function exists f(v,I,z,t) such that for v – u(f(v,I,z,t)), v = 1 when the function is greater than zero.

Using this level as a 'logical' voltage, its 'logical' inverse is v = 1– u(f(v,I,z,t)), which is unity when the 'u' argument is NOT greater than zero.

The expression for the 'B' generator could then be expressed as:

V = u(v(n1,0)–VF)*VF +
(1 – u(–VR–v(n1,0)))* (1– u(v(n1,0)–VF))*v(n1,0) +
u(–VR–v(n1,0))*–VR

The first line is the region 1 condition. If v(n1,0), the input voltage, is greater than VF, the device is conducting with a voltage drop of VF from anode to cathode.

The second line is the region 2 condition. In this case if the input level is NOT more negative than -VR (the breakdown voltage is specified as a positive level) AND simultaneously, VIN is NOT more positive than VF (the input level is less than VF) THEN the voltage is equal to v(n1,0), the input voltage. The device is essentially non-conducting in this region.

The third line is the region 3 condition. If the input voltage is more negative than -VR (making the 'u' function argument positive), then the terminal voltage is equal to -VR. The device is conducting in the reverse, breakdown region with a voltage drop of -VR from anode to cathode.

A test circuit incorporating this 3 region model with some modification is shown in Figure 6 following: The difference in this modified test circuit is that the three regions are separately modeled with individual generators for each region, and no modifying voltage terms for the three regions.

The purpose of this test circuit is to ensure that the each individual modifier covers the appropriate region, and sum of all three of the modifiers is unity, showing that all regions are accounted for in the three terms. VIN is set to 10 VAC, VF is set to 2 volts (to accentuate the conduction region) and VR is set to 8 volts. In the circuit shown below, the generator B1 (voltage v(n3)) produces the multiplier for the VF term, region 1, B3 (voltage v(n5)) produces the multiplier for the VR term, region 3, and generator B2 (voltage v(n4)) produces the multiplier for the v(n1,0) term, region 2. B4 is used as a test, summing the voltages from the B1 through B3 generators. B4 should always produce a unity output after the simulation is started, showing that all regions are covered, and that there are no gaps in coverage, and also that there is no overlap (the sum never exceeds unity).

The graph of the output is shown in Figure.

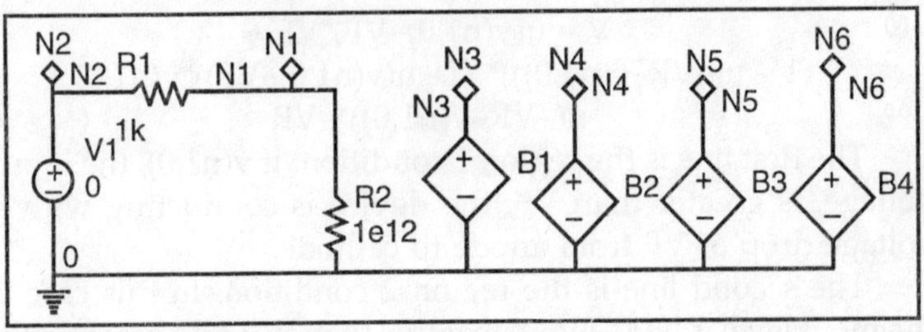

Fig. Perfect Diode Region Test Circuit Graph

It is difficult to see the details clearly in Figure above, however, creating the circuit and running the simulation shows that the results are as desired.

The next step is to modify the circuit of Figure 6 slightly to add the multiplier coefficients, and to place generator B4 (in series with a very small resistor, which is required). Also, the large resistor is deleted. This is performed in the schematic of Figure below. Note again that VF is set to 2v, and VR to 8v, with VIN being a 10V peak sine wave.

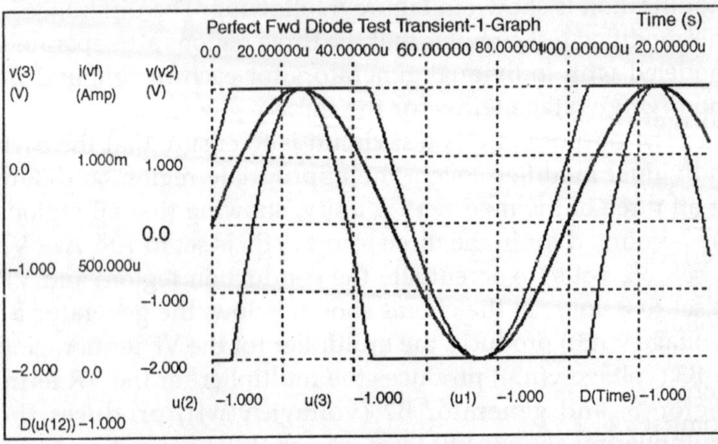

Fig. Preliminary Perfect Diode Test Circuit

A transient analysis graph of the output is as shown in Figure below:

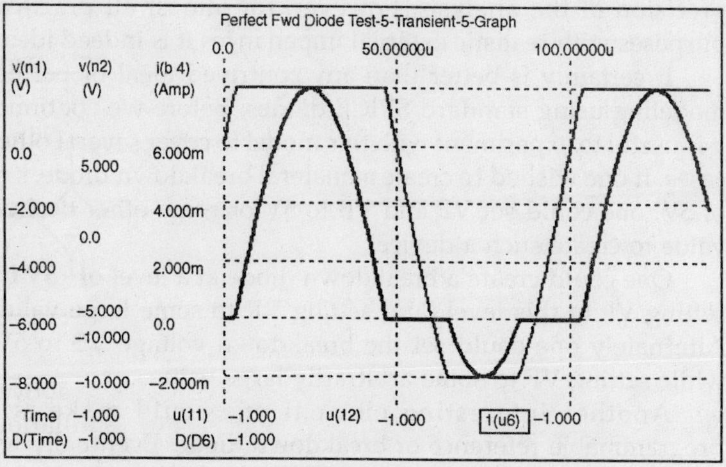

Fig. Preliminary Perfect Diode Test Circuit Graph

The output may seem a little strange at first glance, but some thought shows that it is indeed correct. During the positive half cycle of applied voltage, when the voltage applied is greater than the selected VF of 2V, current will flow. The peak current will be 8mA.

The voltage across this perfect diode model will then be 2V. During the negative half cycle of the applied voltage, when the voltage is less than -8V, the selected value of VB being 8V, current will flow in the opposite sense to that of the forward bias condition. In this case, it will be 2mA peak.

During the interval when the applied voltage level is NOT greater than VF, and simultaneously NOT less than -VB, the perfect diode model will have a voltage equal to and opposing the input voltage, creating an essentially open circuit to that voltage.

The graph SEEMS to show a linear change of voltage across the diode model during this interval, however, by comparing the input voltage to the voltage across the diode model one can see that they are indeed equal, hence no current is flowing.

Actually, there is some slight error due to the programme using a Gmin value of conductance across each element, as well as computational rounding of values and numerical precision of the analysis, however, for almost all practical purposes with realistic external impedances it is indeed ideal.

It certainly is better than any contrived ideal or perfect modeling using standard SPICE diodes. Before we continue, it is useful to experiment with this model to cover several other cases. If one wished to create a bilateral breakdown diode, say of 5V, one could set VF and VB to 5V or some other desired value to create such a device.

One could create a breakdown diode at a level of +5V by setting VF to this level (and setting VR to some large value. Alternately one could set the breakdown voltage VB to 5V, while setting VF to some arbitrarily large value.

Another interesting element one could make is a programmable reference or breakdown diode. Eventually we will create a parameterized subcircuit model for this device, and VR and VF will be passed values.

These values could be variable in value, such that the conduction characteristics could dynamically controlled.

Note that VR and VF could be zero values.

The last step in testing the model is to merge the three condition generators into a single generator.

This is shown in Figure:

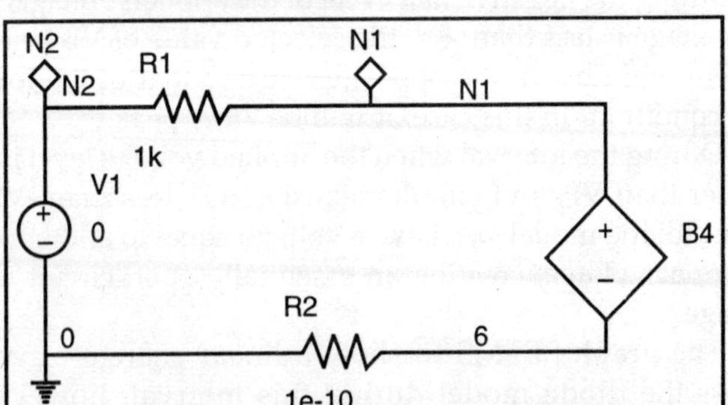

Fig. Final Perfect Diode Test Mode

The graph for this test circuit is shown in Figure:

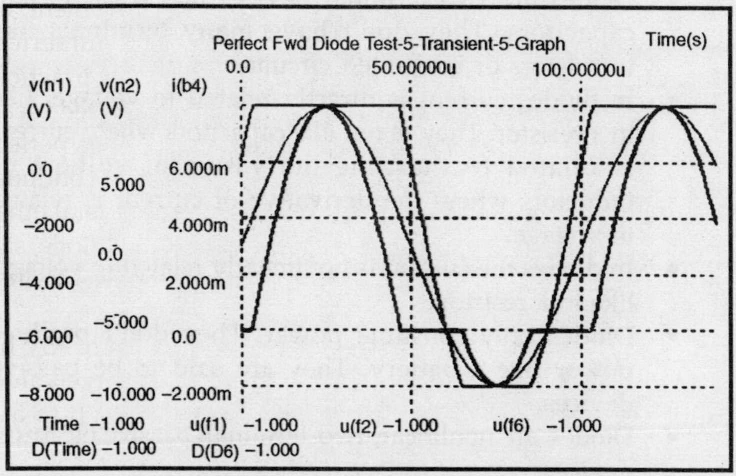

Fig. Perfect Diode Final Test Circuit Graph

The graph results are as expected, with no surprises. Examination of the graph will show that the N2 voltage tracks the input voltage in the non-conduction region, despite the seemingly linear nature of the N2 voltage.

The circuit is now ready to be converted into a parameterized subcircuit, using passed values of VF and VR. A schematic of such a circuit is shown in Figure below:

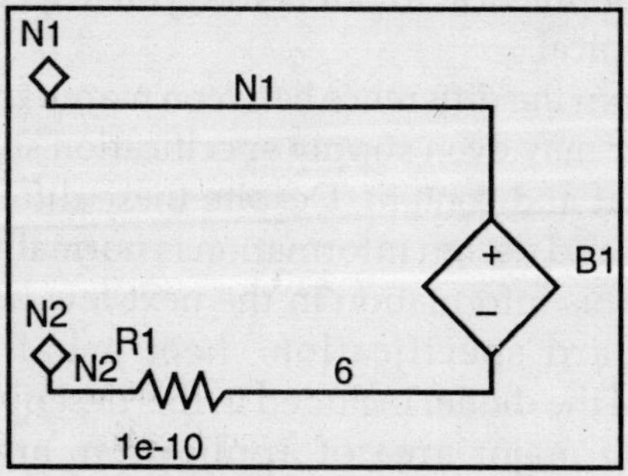

Fig. Perfect Diode Final Model Schematic

CHARACTERISTICS OF DIODE

Diodes have the following characteristics:

- Diodes are two terminal devices like resistors and capacitors. They don't have many terminals like transistors or integrated circuits.
- In diodes current is directly related to voltage, like in a resistor. They're not like capacitors where current is related to the time derivative of voltage or inductors where the derivative of current is related to voltage.
- In diodes the current is not linearly related to voltage, like in a resistor.
- Diodes only consume power. They don't produce power like a battery. They are said to be passive devices.
- Diodes are nonlinear, two terminal, passive electrical devices.

Semiconductor diodes have properties that enable them to perform many different electronic functions. To do their jobs, engineers and technicians must be supplied with data on these different types of diodes. The information presented for this purpose is called DIODE CHARACTERISTICS.

These characteristics are supplied by manufacturers either in their manuals or on specification sheets (data sheets). Because of the scores of manufacturers and numerous diode types, it is not practical to put before you a specification sheet and call it typical.

Aside from the difference between manufacturers, a single manufacturer may even supply specification sheets that differ both in format and content. Despite these differences, certain performance and design information is normally required. We will discuss this information in the next few paragraphs.

A standard specification sheet usually has a brief description of the diode. Included in this description is the type of diode, the major area of application, and any special features. Of particular interest is the specific application for which the diode is suited.

The manufacturer also provides a drawing of the diode which gives dimension, weight, and, if appropriate, any identification marks. In addition to the above data, the following information is also provided: a static operating table (giving spot values of parameters under fixed conditions), sometimes a characteristic curve similar to the one in figure 1-20 (showing how parameters vary over the full operating range), and diode ratings (which are the limiting values of operating conditions outside which could cause diode damage).

In general, diodes tend to permit current flow in one direction, but tend to inhibit current flow in the opposite direction. The graph below shows how current can depend upon voltage for a diode.

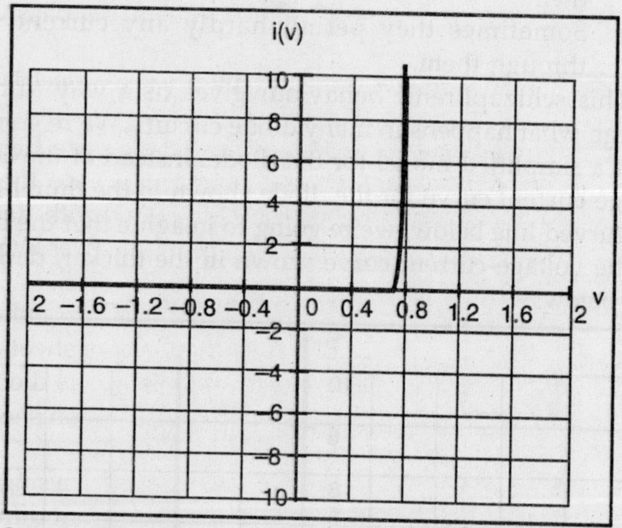

Note the following:

- When the voltage across the diode is positive, a lot of current can flow once the voltage becomes large enough.
- When the voltage across the diode is negative, virtually no current flows.

The circuit symbol for a diode is designed to remind you that current flows easily through a diode in one direction. The circuit symbol for a diode is shown below together with common conventions for current through the diode and voltage across the diode.

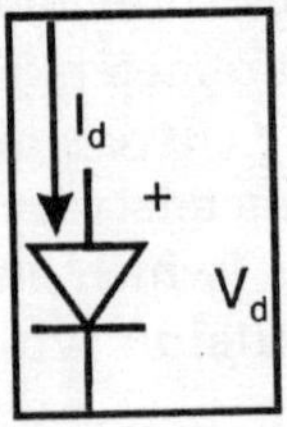

ABOUT DIODES

Diodes are a little schizophrenic:

- Sometimes they let a lot of current flow through them,
- Sometimes they permit hardly any current flow through them.

This schizophrenic behaviour gives us a way to think through what happens in many diode circuits. We're going to adopt a simplified model for the diode. Instead of the actual voltage-current curve for the diode shown in the thin, lighter red, curved line below, we're going to imagine that the diode has the voltage-current curve shown in the thicker, dark red lines below.

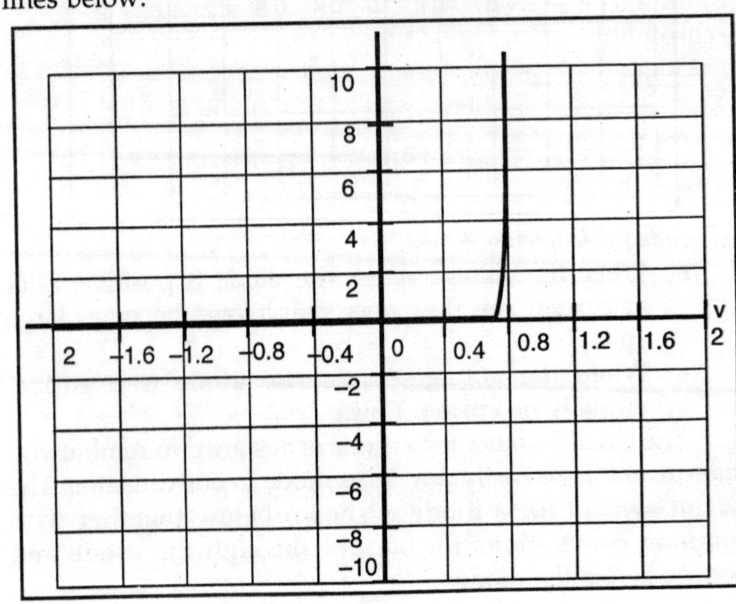

The approximate voltage-current curve gives us one way to analyse circuits that contain diodes, and to account for their schizophrenic behaviour.

- When current is flowing, this approximate model predicts no voltage across the diode. In this situation, we say that the diode isON.
- When the voltage across the diode is negative, this approximate model predicts no current flowing through the diode. In this situation, we say that the diode is OFF.

Now, consider this kind of simplified model for the diode.

- When the diode is ON, it has no voltage across it so it acts like a short circuit! When the diode is ON, the current through the diode is positive, and the voltage across the diode is zero.
- When the diode is OFF, current is zero, so it acts like an open circuit! When the diode is OFF, the voltage across the diode is negative, and the current through the diode is zero.

So, this idealized model for the diode is sometimes an open circuit, and somtimes a closed circuit - truly schizophrenic! This model for the diode is often referred to as the ideal diode model.

Chapter 3

Rectifying Circuits

LOAD LINE ANALYSIS OF DIODE CIRCUIT

INTRODUCTION

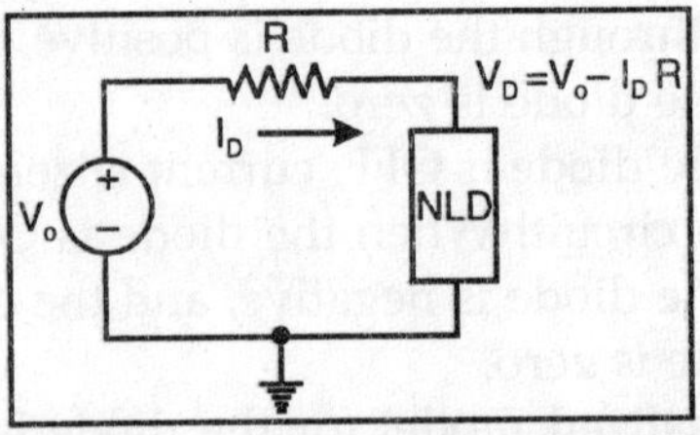

Consider the above ciruit, where NLD is a nonlinear device (e.g., a diode) which has a known IV characteristic. Since V_D must satisfy both the known IV characteristic and the equation,

$$V_D = V_0 - I_D R$$

both may be plotted on the same graph as in the following diode example:

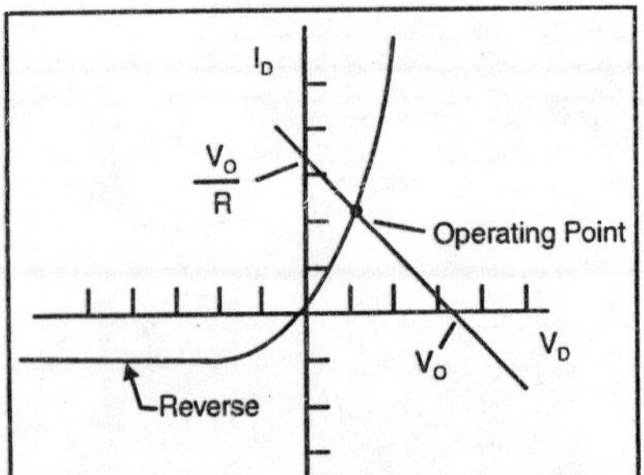

Fig. Forward Biased

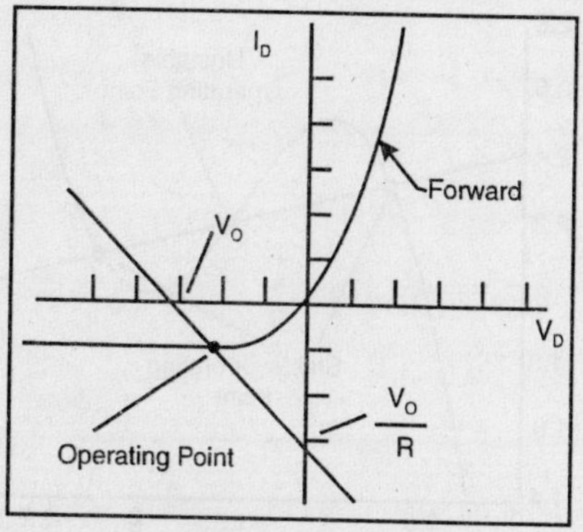

Fig. Reverse Biased

Another interesting example of a NLD is the (Esaki) tunnel diode which has a region of "negative" resistance in its characteristic.

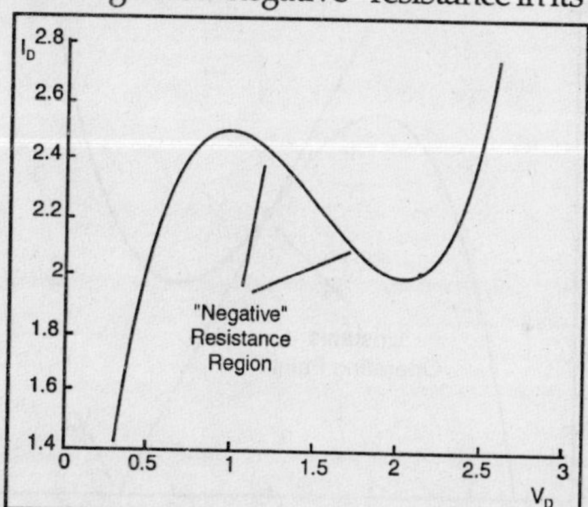

Fig. Characteristic of an Esaki (Tunnel) Diode

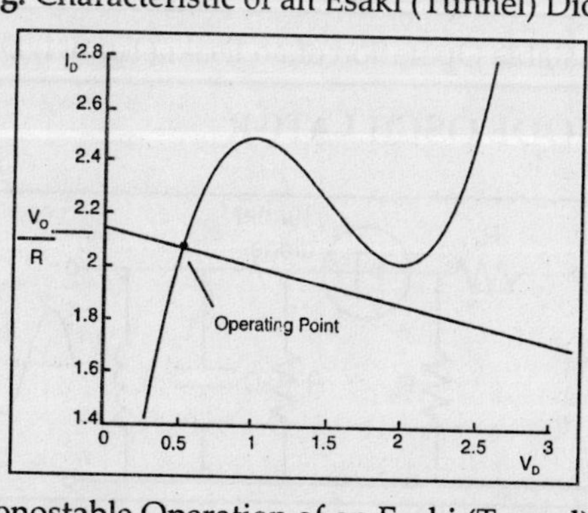

Fig. Monostable Operation of an Esaki (Tunnel) Diode

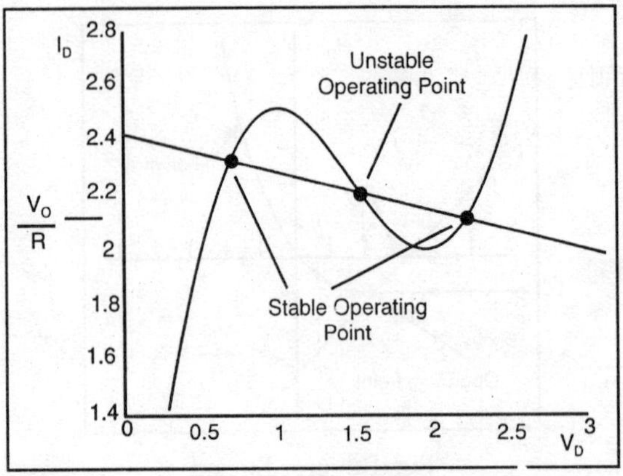

Fig. Bistable Operation of an Esaki (Tunnel) Diode

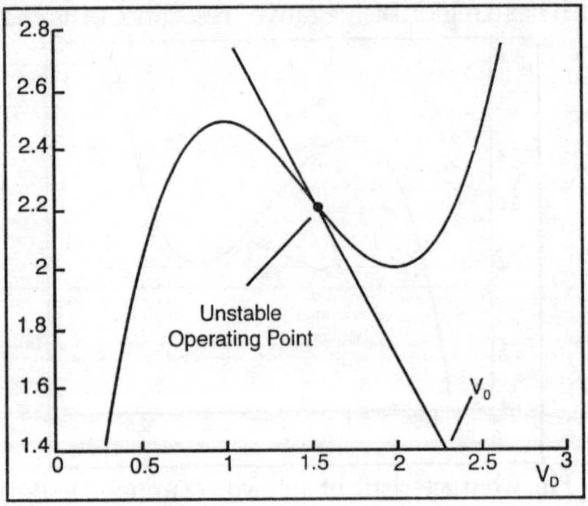

Fig. Astable Operation of an Esaki (Tunnel) Diode

TUNNEL DIODE OSCILLATOR

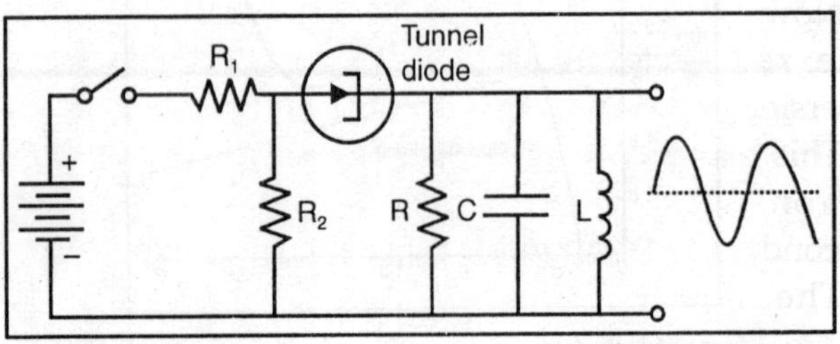

The excess small forward current flow in a tunnel or Esaki diode is a manifestation of quantum mechanical tunneling. Consider the following pictures:

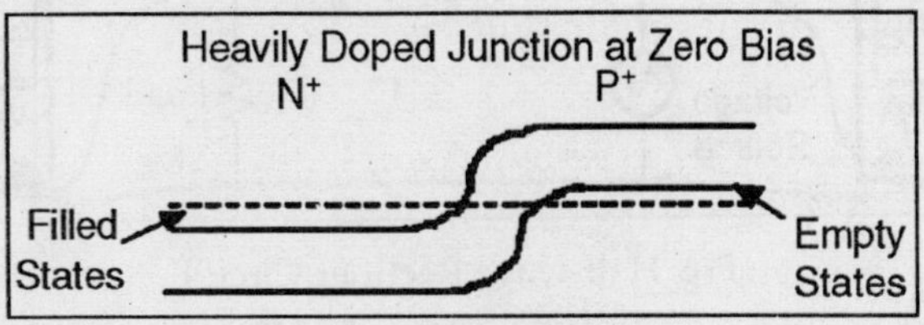

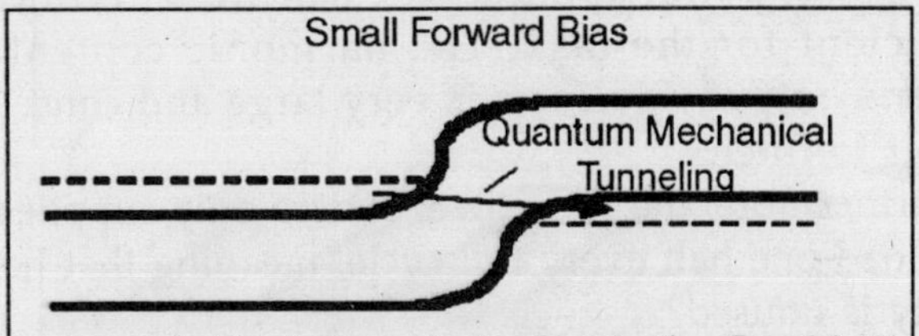

Completely Filled States in the Conduction Band (n^{++}) Accross from Empty States in the Valence band (b^{++}) for a Limited Range of Applied Bias!!!

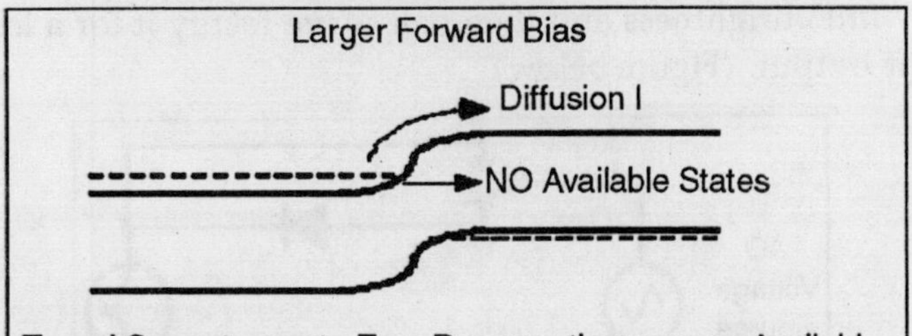

Tunnel Current goes to Zero Because there are no Available States for the Electrons to Tunnel into and Thus only diffusion or G-R Current can occur.

THE HALF WAVE RECTIFIER

Now we come to the most popular application of the diode: *rectification*. Simply defined, rectification is the conversion of alternating current (AC) to direct current (DC).

This involves a device that only allows one-way flow of electrons. As we have seen, this is exactly what a semiconductor diode does.

The simplest kind of rectifier circuit is the *half-*

wave rectifier. It only allows one half of an AC waveform to pass through to the load. (Figure below)

Fig. Half-wave Rectifier Circuit.

For most power applications, half-wave rectification is insufficient for the task. The harmonic content of the rectifier's output waveform is very large and consequently difficult to filter.

Furthermore, the AC power source only supplies power to the load one half every full cycle, meaning that half of its capacity is unused.

Half-wave rectification is, however, a very simple way to reduce power to a resistive load. Some two-position lamp dimmer switches apply full AC power to the lamp filament for "full" brightness and then half-wave rectify it for a lesser light output. (Figure below)

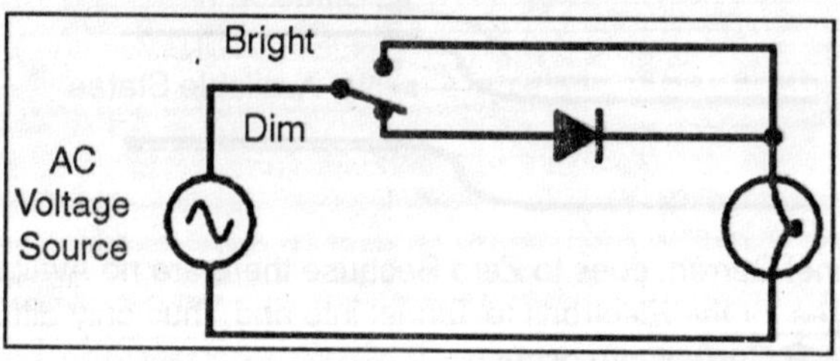

Fig. Half-wave Rectifier Application: Two Level Lamp Dimmer.

In the "Dim" switch position, the incandescent lamp receives approximately one-half the power it would normally receive operating on full-wave AC. Because the half-wave rectified power pulses far more rapidly than the filament has time to heat up and cool down, the lamp does not blink.

Instead, its filament merely operates at a lesser temperature than normal, providing less light output. This principle of "pulsing" power rapidly to a slow-responding load device to control the electrical power sent to it is common in the world of industrial electronics.

Since the controlling device (the diode, in this case) is either fully conducting or fully nonconducting at any given time, it dissipates little heat energy while controlling load power, making this method of power control very energy-efficient.

This circuit is perhaps the crudest possible method of pulsing power to a load, but it suffices as a proof-of-concept application.

VOLTAGE REGULATOR

A Zener Diode is an electronic component which can be used to make a very simple voltage regulator circuit. This circuit enables a fixed stable voltage to be taken from an unstable voltage source such as thebattery bank of a renewable energy system which will fluctuate depending on the state of charge of the bank.

Zener Diode Voltage Regulator Circuit:

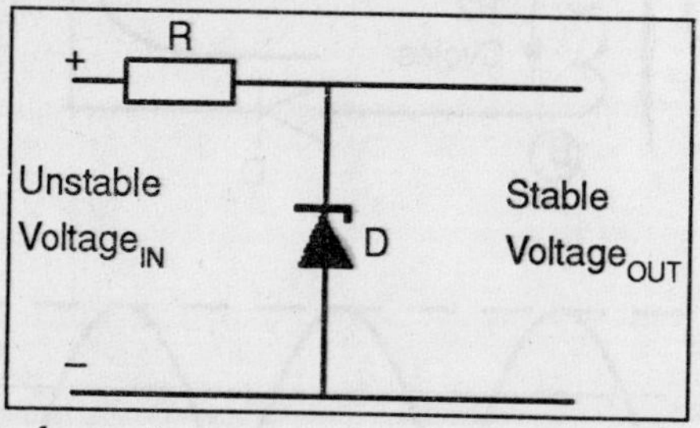

Pictured above is a very simple voltage regulator circuit requiring just one zener diode (available from the REUK Shop) and one resistor.

As long as the input voltage is a few volts more than the desired output voltage, the voltage across the zener diode will be stable.

As the input voltage increases the current through the Zener diode increases but the voltage drop remains constant - a feature of zener diodes.

Therefore since the current in the circuit has increased the voltage drop across the resistor increases by an amount equal to the difference between the input voltage and the zener voltage of the diode.

FULL WAVE RECTIFIER

In a Full-wave rectifier circuit two diodes are now used, together with a transformer whose secondary winding is split equally into two and has a common centre tapped connection, (C). Now each diode conducts in turn when its Anode terminal is positive with respect to the centre point C as shown below.

FULL-WAVE RECTIFIER CIRCUIT

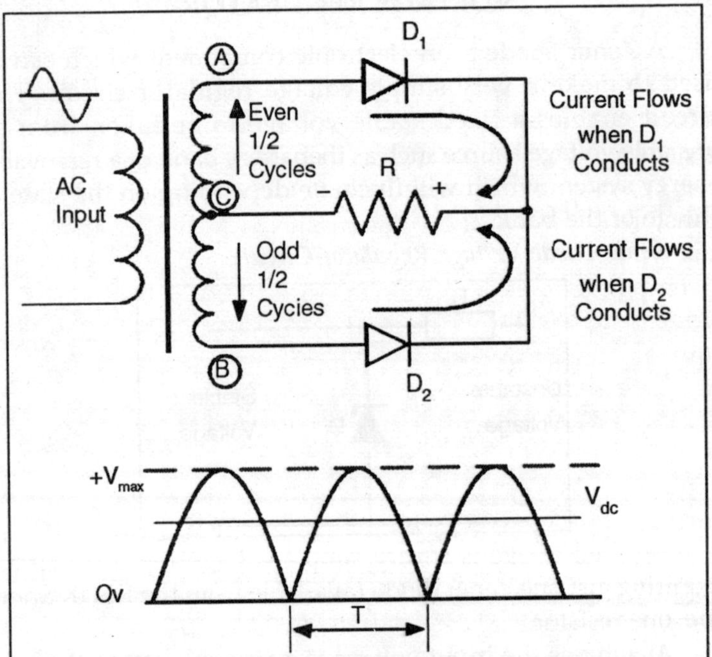

Resultant Output Waveform

The circuit consists of two *Half-wave* rectifiers connected to a single load resistance with each diode taking it in turn to supply current to the load.

When point A is positive with respect to point B, diode D_1 conducts in the forward direction as indicated by the arrows. When point B is positive (in the negative half of the cycle) with respect to point A, diode D_2 conducts in the forward direction and the current flowing through resistor R is in the same direction for both circuits.

As the output voltage across the resistor R is the sum of the two waveforms, this type of circuit is also known as a "bi-phase" circuit.

As the spaces between each half-wave developed by each diode is now being filled in by the other diode the average DC output voltage across the load resistor is now double that of the single half-wave rectifier circuit and is about $0.637V_{max}$ of the peak voltage, assuming no losses.

$$V_{d.c.} = \frac{V_{max}}{\pi} = 0.637V_{max} = 0.9V_S$$

The peak voltage of the output waveform is the same as before for the half-wave rectifier provided each half of the transformer windings have the same rms voltage value.

To obtain a different d.c. voltage output different transformer ratios can be used, but one main disadvantage of this type of rectifier is that having a larger transformer for a given power output with two separate windings makes this type of circuit costly compared to a "Bridge Rectifier" circuit equivalent.

THE BRIDGE RECTIFIER

Another type of circuit that produces the same output as a full-wave rectifier is that of the Bridge Rectifier. This type of single phase rectifier uses 4 individual rectifying diodes connected in a "bridged" configuration to produce the desired output but does not require a special centre tapped transformer, thereby reducing its size and cost.

The single secondary winding is connected to one side of the diode bridge network and the load to the other side as shown below.

THE DIODE BRIDGE RECTIFIER

The 4 diodes labelled D_1 to D_4 are arranged in "series pairs" with only two diodes conducting current during each half cycle. During the positive half cycle of the supply, diodes D1 and D2 conduct in series while diodes D3 and D4 are reverse biased and the current flows through the load as shown below.

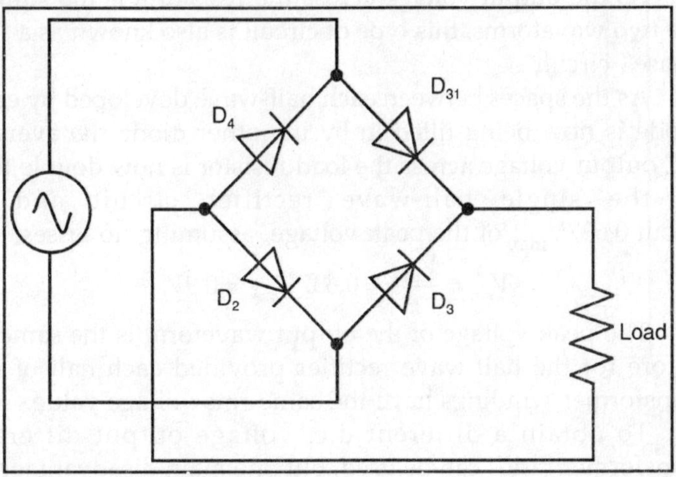

The Positive Half-cycle

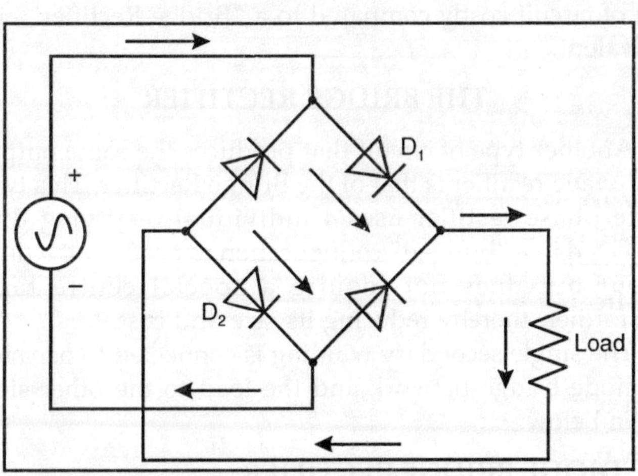

During the negative half cycle of the supply, diodes D3 and D4 conduct in series, but diodes D1 and D2 switch of as they are now reverse biased. The current flowing through the load is the same direction as before.

The Negative Half-cycle

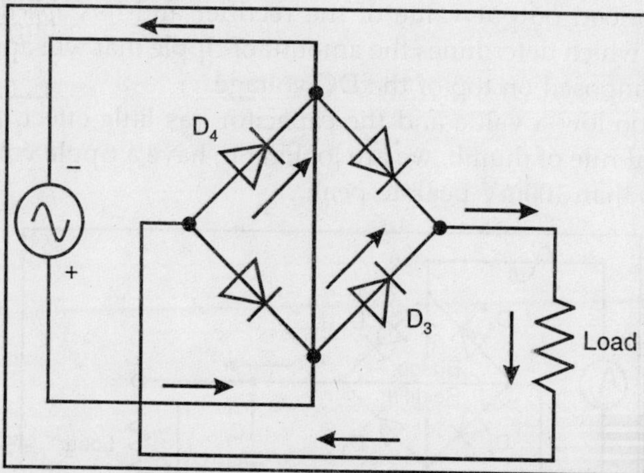

As the current flowing through the load is unidirectional, so the voltage developed across the load is also unidirectional the same as for the previous two diode full-wave rectifier, therefore the average DC voltage across the load is $0.637V_{max}$ and the ripple frequency is now twice the supply frequency (e.g. 100Hz for a 50Hz supply).

SMOOTHING CAPACITOR

We saw in the previous section that the single phase half-wave rectifier produces an output wave every half cycle and that it was not practical to use this type of circuit to produce a steady DC supply. The full-wave bridge rectifier however, gives us a greater mean DC value (0.637 Vmax) with less superimposed ripple while the output waveform is twice that of the frequency of the input supply frequency.

We can therefore increase its average DC output level even higher by connecting a suitable smoothing capacitor across the output of the bridge circuit as shown below.

Full-wave Rectifier with Smoothing Capacitor

The smoothing capacitor converts the full-wave rippled output of the rectifier into a smooth DC output voltage. Two important parameters to consider when choosing a suitable a capacitor are its *Working Voltage,* which must be higher than the no-load output value of the rectifier and its *Capacitance Value,* which determines the amount of ripple that will appear superimposed on top of the DC voltage.

Too low a value and the capacitor has little effect. As a general rule of thumb, we are looking to have a ripple voltage of less than 100mV peak to peak.

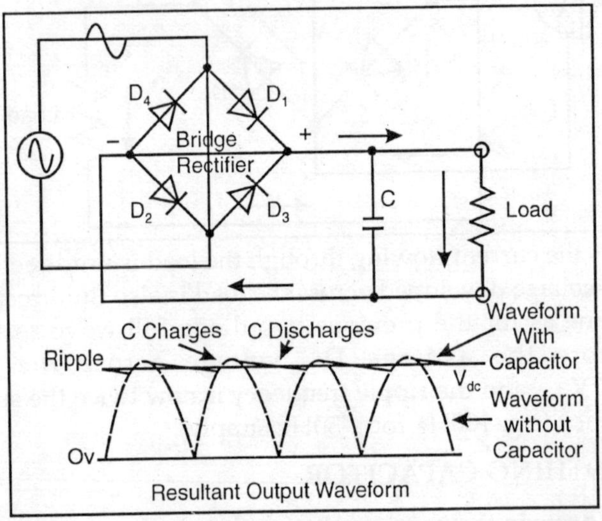

The main advantages of a full-wave bridge rectifier is that it has a smaller AC ripple value for a given load and a smaller reservoir or smoothing capacitor than an equivalent half-wave rectifier. Therefore, the fundamental frequency of the ripple voltage is twice that of the AC supply frequency (100Hz) where for the half-wave rectifier it is exactly equal to the supply frequency (50Hz).

The amount of ripple voltage that is superimposed on top of the DC supply voltage by the diodes can be virtually eliminated by adding a much improved π-filter (pi-filter) to the output terminals of the bridge rectifier. This type of low-pass filter consists of two smoothing capacitors, usually of the same value and a choke or inductance across them to introduce a high impedance path to the alternating ripple component. Another more practical and cheaper alternative is to use a 3-

terminal voltage regulator IC, such as a LM7805 which can reduce the ripple by more than 70dB (Datasheet) while delivering over 1amp of output current.

INDUCTOR FILTER

INDUCTOR FILTERS use an inductor called a choke to filter the pulsating dc input. Because ofthe impedance offered to circuit current, the output of the filter is at a lower amplitude than the input

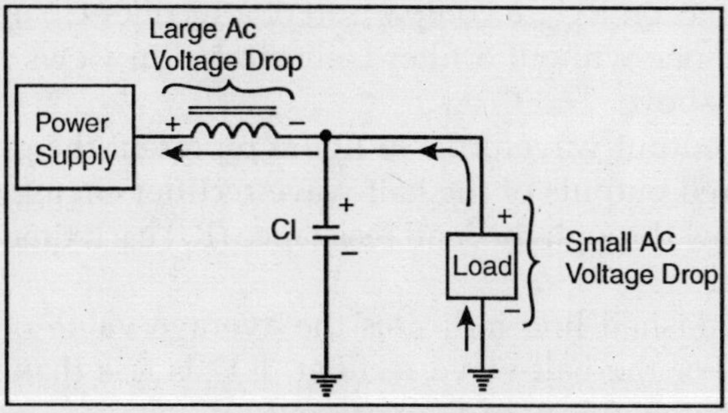

CAPACITOR FILTER

The simple capacitor filter is the most basic type of power supply filter. The application of the simple capacitor filter is very limited. It is sometimes used on extremely high-voltage, low-current power supplies for cathode-ray and similar electron tubes, which require very little load current from the supply. The capacitor filter is also used where the power-supply ripple frequency is not critical; this frequency can be relatively high. The capacitor (C1) shown in figure is a simple filter connected across the output of the rectifier in parallel with the load.

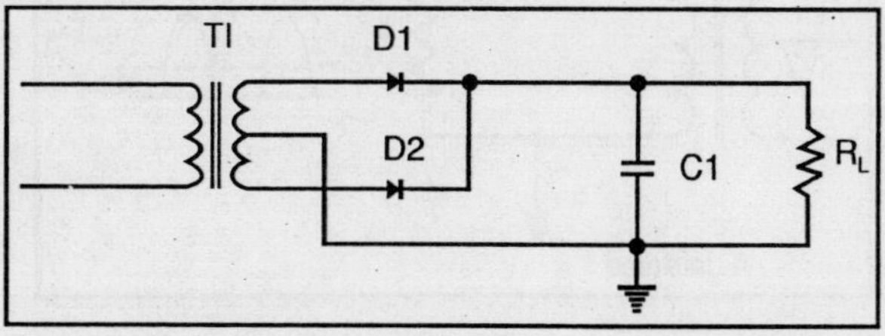

Fig. Full-wave Rectifier with a Capacitor Filter.

When this filter is used, the RC charge time of the filter capacitor (C1) must be short and the RC discharge time must be long to eliminate ripple action.

In other words, the capacitor must charge up fast, preferably with no discharge at all. Better filtering also results when the input frequency is high; therefore, the full-wave rectifier output is easier to filter than that of the half-wave rectifier because of its higher frequency.

For you to have a better understanding of the effect that filtering has on E_{avg}, a comparison of a rectifier circuit with a filter and one without a filter is illustrated in views A and B of figure above.

The output waveforms in figure represent the unfiltered and filtered outputs of the half-wave rectifier circuit. Current pulses flow through the load resistance (R_L) each time a diode conducts.

The dashed line indicates the average value of output voltage. For the half-wave rectifier, E_{avg} is less than half (or approximately 0.318) of the peak output voltage.

This value is still much less than that of the applied voltage. With no capacitor connected across the output of the rectifier circuit, the waveform in view A has a large pulsating component (ripple) compared with the average or dc component.

When a capacitor is connected across the output (view B), the average value of output voltage (E_{avg}) is increased due to the filtering action of capacitor C1.

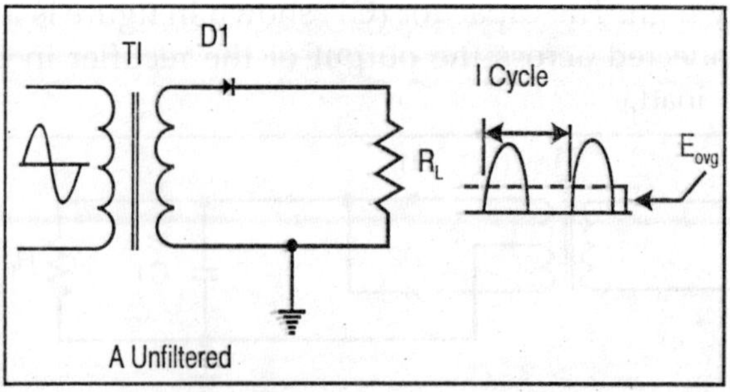

Fig. A. Half-wave Rectifier with and without Filtering.

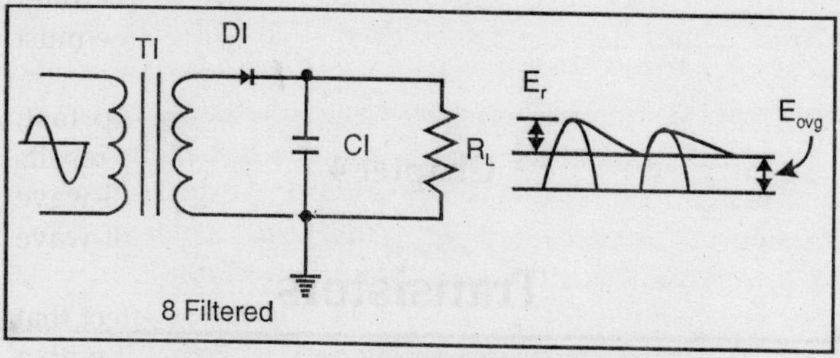

Fig. B. Half-wave Rectifier with and without Filtering.

The value of the capacitor is fairly large (several microfarads), thus it presents a relatively low reactance to the pulsating current and it stores a substantial charge.

RIPPLE FACTOR

Ripple factor can be defined as the variation of the amplitude of DC (Direct current) due to improper filtering of AC power supply.

It can be measured by,

$$RF = v_{rms} / v_{dc}$$

Chapter 4

Transistors

METAL-SEMICONDUCTOR-FIELD-EFFECT-TRANSISTORS (MESFET)

INTRODUCTION

The Metal-Semiconductor-Field-Effect-Transistor (MESFET) consists of a conducting channel positioned between a source and drain contact region as shown in the figure below. The carrier flow from source to drain is controlled by a Schottky metal gate. The control of the channel is obtained by varying the depletion layer width underneath the metal contact which modulates the thickness of the conducting channel and thereby the current.

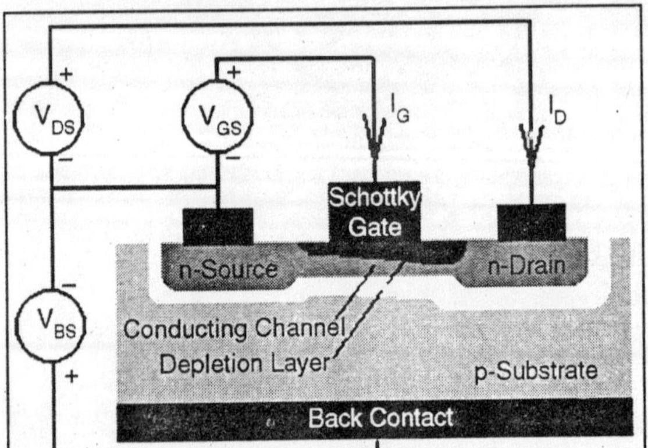

The key advantage of the MESFET is the higher mobility of the carriers in the channel as compared to the MOSFET.

Since the carriers located in the inversion layer of a MOSFET have a wave function which extends into the oxide, their mobility - also referred to as surface mobility - is less than half of the mobility of bulk material. As the depletion region separates the carriers from the surface their mobility is close to that of bulk material. The higher mobility leads to a higher current, trans conductance and transit frequency of the device.

The disadvantage of the MESFET structure is the presence of the Schottky metal gate. It limits the forward bias voltage on the gate to the turn-on voltage of the Schottky diode.

This turn-on voltage is typically 0.7 V for GaAs Schottky diodes. The threshold voltage therefore must be lower than this turn-on voltage. As a result it is more difficult to fabricate circuits containing a large number of enhancement-mode MESFET. The higher transit frequency of the MESFET makes it particularly of interest for microwave circuits.

While the advantage of the MESFET provides a superior microwave amplifier or circuit, the limitation by the diode turn-on is easily tolerated. Typically depletion-mode devices are used since they provide a larger current and larger trans conductance and the circuits contain only a few transistors, so that threshold control is not a limiting factor. The buried channel also yields a better noise performance as trapping and release of carriers into and from surface states and defects is eliminated.

The use of GaAs rather than silicon MESFETs provides two more significant advantages: first of all the room temperature mobility is more than 5 times larger, while the saturation velocity is about twice that in silicon. Second it is possible to fabricate semi-insulating (SI) GaAs substrates, which eliminates the problem of absorbing microwave power in the substrate due to free carrier absorption.

METAL OXIDE SEMICONDUCTOR FIELD EFFECT TRANSISTOR (MOSFET)

INTRODUCTION

Mosfets are 4-terminal devices with a gate that controls

the conduction of the drain to source channel. The body (or substrate) of the typical n-channel enhancement mosfet is a P-type material with the drain and source formed from N-type materials.

The oxide layer insulates the metal gate connection from the other layers. There are holes in the oxide layer left for the metallic contacts of the drain and source.

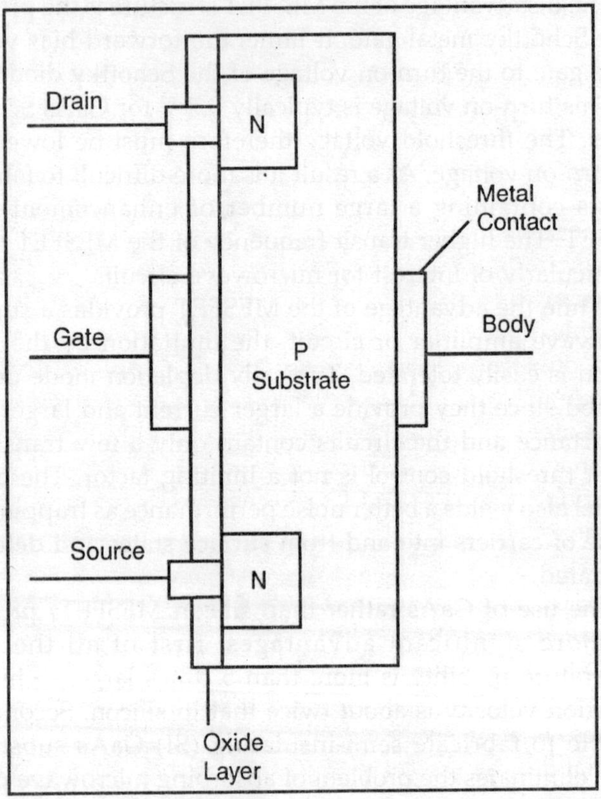

The metal oxide semiconductor field effect transistor (MOSFET) is one of the cornerstones of modern semiconductor technology.

The general structure is a lightly doped *p*-type substrate, into which two regions, the source and the drain, both of heavily doped *n*-type semiconductors have been embedded. The symbol *n+* is used to denote this heavy doping.

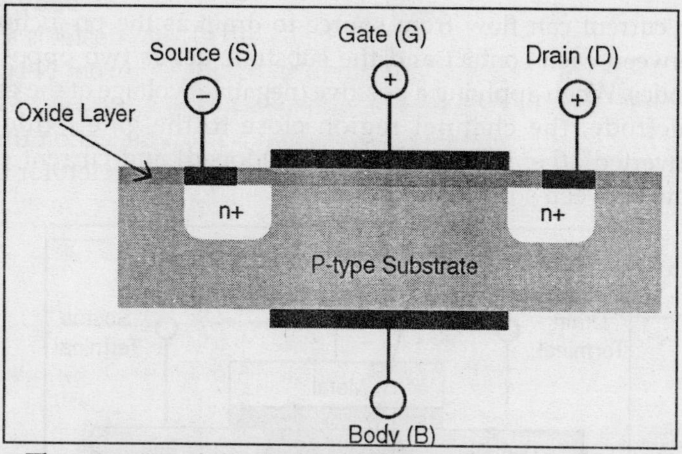

The source and the drain are about 1 ìm apart. Metallized contacts are made to both source and drain, generally using aluminium. The rest of the substrate surface is covered with a thin oxide film, typically about 0.05 ìm thick. The gate electrode is laid on top of the insulating oxide layer, and the body electrode in the above diagram provides a counter electrode to the Gate. The thin oxide film contains silicon dioxide (SiO_2), but it may well also contain silicon nitride (Si_3N_4) and silicon oxynitride (Si_2N_2O).

The *p*-type doped substrate is only very lightly doped, and so it has a very high electrical resistance, and current cannot pass between the source and drain if there is zero voltage on the gate. Application of a positive potential to the gate electrode creates a strong electric field across the *p*-type material even for relatively small voltages, as the device thickness is very small and the field strength is given by the potential difference divided by the separation of the gate and body electrodes.

CONSTRUCTION

A MOSFET consists of an n-(p-)doped silicon substrate with two, highly p-(n-)doped contacts, source and drain. The so-called channel region inbetween is covered by an insulating layer, the gate-oxide, which is in contact with the gate electrode. Without applying a voltage at the gate electrode, no current can flow from source to drain as the pn-juctions between each contact and the substrate act as two opposite diodes. When applying a positive (negative) voltage at the gate electrode, the channel region close to the gate oxide is "inverted" (i.e. from n-(p-) to p-(n-)doped) and current can flow between source and drain.

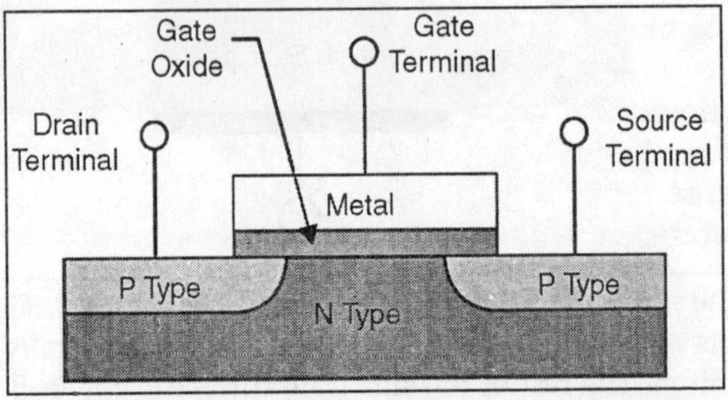

Central to the functionality is the thin insulating layer, the gate-oxide. The gate-oxide acts as the dielectricum of a capacitor which attracts charge carriers into the channel region. Up to now, silicon dioxide (SiO_2) has been used as a gate-oxide. The so far unparallelized electric and structural properties of silicon's native oxide - commonly known as window glass - are said to be one of the main reasons that silicon is today's semiconductor of choice.

Due to electrical considerations, the gate oxide thickness has to be proportional to the gate length (i.e. the distance between source and drain). The lateral scaling of transistor dimensions thus directly translates into a reduction of the gate-oxide thickness. Right now, oxide with a thickness of about 1.5 nm (one fifty-thousandth of the width of human hair) are employed.

In 2007 less than 1 nm will be needed which corresponds to only 4 atomic layers. At these thick nesses, quantum mechanical tunneling currents through the oxide become intolerable.

WORKING

N-P junctions act as diodes, and because the layers of the

mosfet have P-type and N-type materials in close contact, diodes are formed from the body to the drain or source, as shown in the drawing here of the 4-terminal mosfet. If you were to leave the body connection floating, there could be no voltage flow through the two PN body diodes because they are reverse connected to each other.

However, in the 3-terminal mosfets (like the BS170) that are commonly used in guitar effects pedals, the body and source are connected internally and this shorts out the PN junction of the lower diode and removes it from the circuit. The top body diode can now conduct from drain to source because of the internal connection and therefore has to be taken into consideration during circuit design.

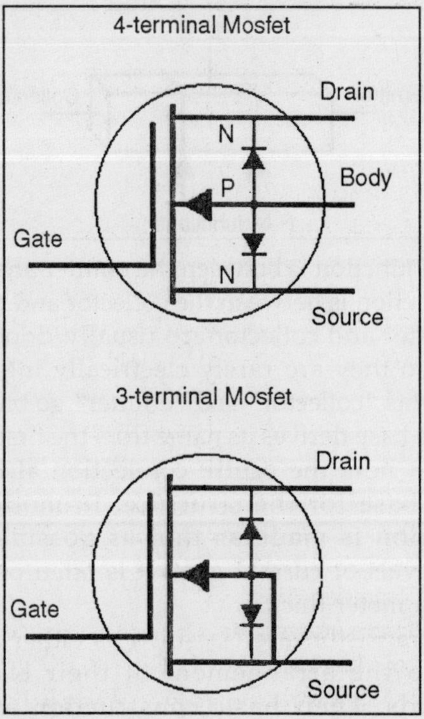

This means that every 3-terminal mosfet has a body diode due the construction of the device. It is not added as a separate component inside the mosfet.

However, there is something even more interesting in the 4-terminal mosfet. The drain-to-body-to-source structure is an N-P-N layered structure... NPN, as in transistor! There is a bipolar transistor in every mosfet!

BIPOLAR JUNCTION TRANSISTORS

A Bipolar Junction Transistor (a BJT or Bipolar Transistor) is an active semiconductor device formed by two P-N junctions whose function is amplification of an electric current.

Bipolar transistors are made from 3 sections of semiconductor material (alternating P-type and N-type), with 2 resulting P-N junctions. Schematically, a bipolar transistor can be thought of in this fashion.

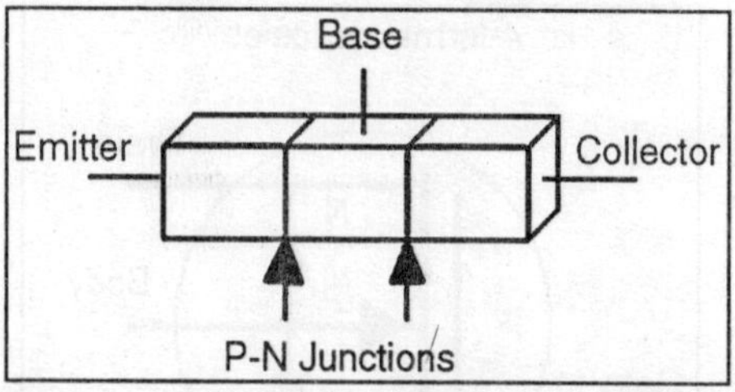

One P-N junction is between the emitter and the base; the other P-N junction is between the collector and the base. Note that the emitter and collector are usually doped somewhat differently, so they are rarely electrically interchangeable. While the terms "collector" and "emitter" go back to vacuum tube days, the base derives its name from the first point-contact transistors – here the centre connection also formed the mechanical base for the structure. In modern practice, the base region is made as thin as possible to achieve reasonable levels of current gain; it is often only about one millionth of a meter thick.

Bipolar transistors are classified as either NPN or PNP according to the arrangement of their N-type and P-type materials. Their basic construction and chemical treatment is implied by their names. So an NPN transistors is formed by introducing a thin region of P-type material between two regions of N-type material.

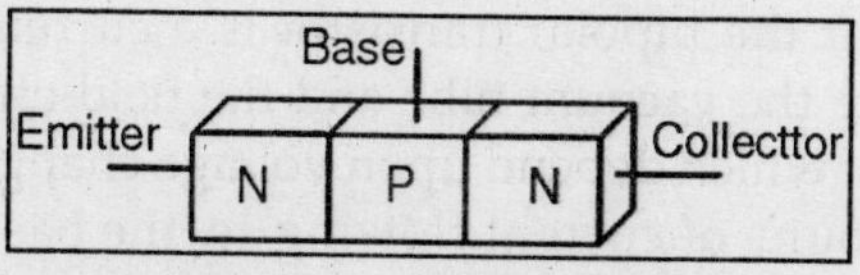

On the other hand, a PNP transistor is formed by introducing a thin region of N-type material between two regions of P-type material.

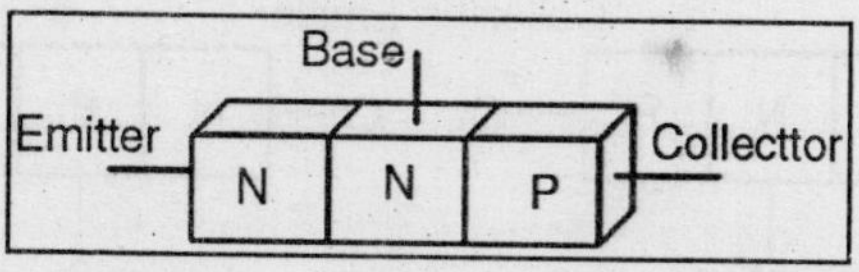

Since the majority and minority current carriers are different for N-type and P-type materials, it stands to reason that the internal operation of the NPN and PNP transistors will also be different. These two basic types of transistors along with their circuit symbols are shown here:

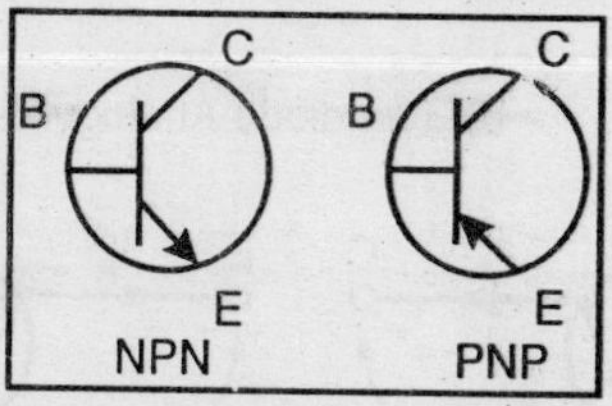

Note that the two symbols are subtly different. The vertical line represents the base (B), the angular line with the arrow on it represents the emitter (E), and the other angular line represents the collector (C). The direction of the arrow on the emitter distinguishes (graphically) the NPN from the PNP transistor. If the arrow points in, (Points iN) the transistor is a PNP. On the other hand if the arrow points out, the transistor is an NPN (Not Pointing iN).

Bear in mind that the arrow always points in the direction of hole flow (current), or from the P-type to N-type sections, no matter whether the P-type section is the emitter or base. On the other hand, electron flow is always "against" the arrow, just like in the junction diode. As a result, a PNP transistor is "triggered" when its base is pulled low; an NPN transistor is "triggered" when its base is brought high.

Note that the bipolar transistor is a current-amplifying device, unlike the vacuum tube and the field-effect transistor (FET), both of which depend upon voltage changes to operate. It is the amount of current flowing in the base circuit that controls the amount of current flowing in the collector circuit.

CONSTRUCTION

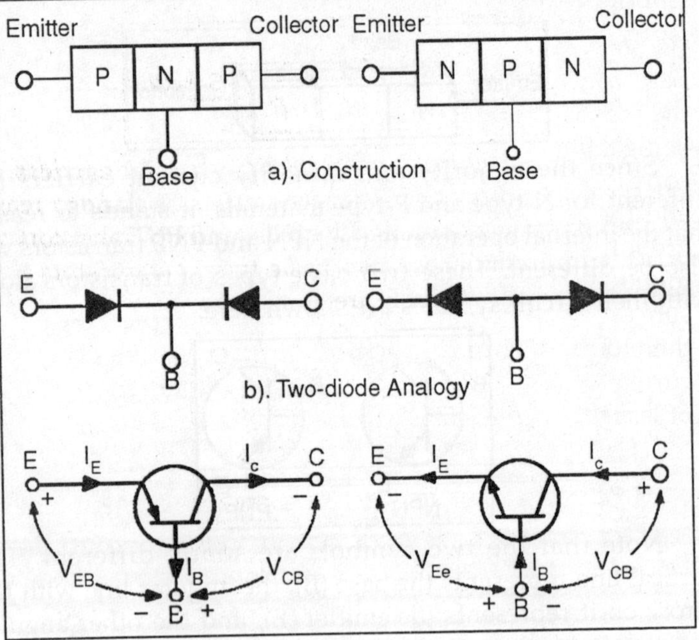

The construction and circuit symbols for both the NPN and PNP bipolar transistor are shown above with the arrow in the circuit symbol always showing the direction of conventional current flow between the base terminal and its emitter terminal, with the direction of the arrow pointing from the positive P-type region to the negative N-type region, exactly the same as for the standard diode symbol.

There are basically three possible ways to connect a Bipolar Transistor within an electronic circuit with each method of connection responding differently to its input signal as the static characteristics of the transistor vary with each circuit arrangement.

- *Common Base Configuration*: Has Voltage Gain but no Current Gain.
- *Common Emitter Configuration*: Has both Current and Voltage Gain.
- *Common Collector Configuration*: Has Current Gain but no Voltage Gain.

Common Base Configuration

As its name suggests, in the Common Base or Grounded Base configuration, the BASE connection is common to both the input signal AND the output signal with the input signal being applied between the base and the emitter terminals. The corresponding output signal is taken from between the base and the collector terminals as shown with the base terminal grounded or connected to a fixed reference voltage point. The input current flowing into the emitter is quite large as its the sum of both the base current and collector current respectively therefore, the collector current output is less than the emitter current input resulting in a Current Gain for this type of circuit of less than "1", or in other words it "Attenuates" the signal.

Common Base Amplifier Circuit

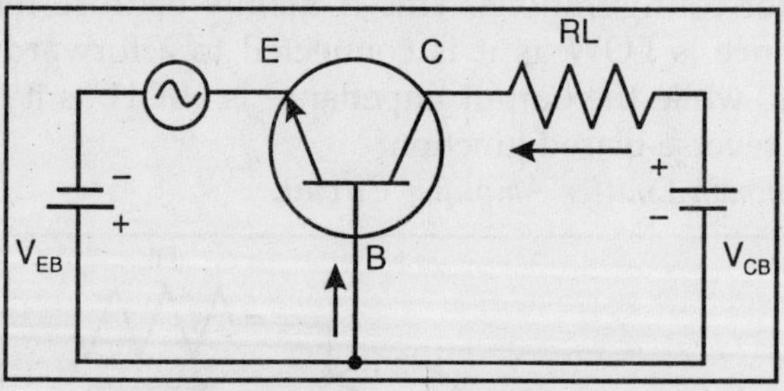

This type of amplifier configuration is a non-inverting voltage amplifier circuit, in that the signal voltages Vin and Vout are In-Phase. This type of arrangement is not very common due to its unusually high voltage gain characteristics. Its Output characteristics represent that of a forward biased diode while the Input characteristics represent that of an illuminated photo-diode. Also this type of configuration has a high ratio of Output to Input resistance or more importantly "Load" resistance (RL) to "Input" resistance (Rin) giving it a value of "Resistance Gain". Then the Voltage Gain for a common base can therefore be given as:

Common Base Voltage Gain:

$$A_V = \frac{I_C \times R_L}{I_e \times R_{IN}} = \alpha \times \frac{R_L}{R_{IN}}$$

The Common Base circuit is generally only used in single stage amplifier circuits such as microphone pre-amplifier or RF radio amplifiers due to its very good high frequency response.

Common Emitter Configuration

In the Common Emitter or Grounded Emitter configuration, the input signal is applied between the base, while the output is taken from between the collector and the emitter as shown.

This type of configuration is the most commonly used circuit for transistor based amplifiers and which represents the "normal" method of connection.

The common emitter amplifier configuration produces the highest current and power gain of all the three bipolar transistor configurations. This is mainly because the input impedance is LOW as it is connected to a forward-biased junction, while the output impedance is HIGH as it is taken from a reverse-biased junction.

Common Emitter Amplifier Circuit:

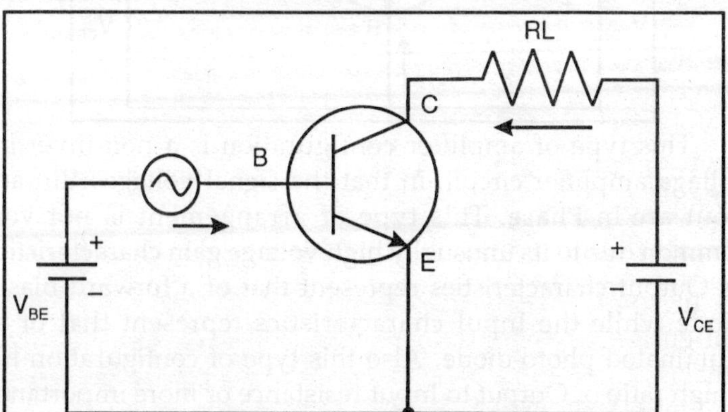

In this type of configuration, the current flowing out of the transistor must be equal to the currents flowing into the transistor as the emitter current is given as Ie = Ic + Ib.

Also, as the load resistance (RL) is connected in series with the collector, the Current gain of the Common Emitter Transistor Amplifier is quite large as it is the ratio of Ic/Ib and is given the symbol of Beta, (β).

Since the relationship between these three currents is determined by the transistor itself, any small change in the base current will result in a large change in the collector current. Then, small changes in base current will thus control the current in the Emitter/Collector circuit.

By combining the expressions for both Alpha, α and Beta, β the mathematical relationship between these parameters and therefore the current gain of the amplifier can be given as:

$$I_E = I_C + I_B$$

$$\alpha = \frac{I_C}{I_C} and\, \beta = \frac{I_C}{I_B}$$

$$\alpha = \frac{\beta}{\beta + 1} \beta = \frac{\alpha}{1 - \alpha}$$

Where: "Ic" is the current flowing into the collector terminal, "Ib" is the current flowing into the base terminal and "Ie" is the current flowing out of the emitter terminal.

Then to summarise, this type of bipolar transistor configuration has a greater input impedance, Current and Power gain than that of the common Base configuration but its Voltage gain is much lower. The common emitter is an inverting amplifier circuit resulting in the output signal being 180° out of phase with the input voltage signal.

Common Collector Configuration

In the Common Collector or Grounded Collector configuration, the collector is now common and the input signal is connected to the Base, while the output is taken from the Emitter load as shown. This type of configuration is commonly known as a Voltage Follower or Emitter Follower circuit.

The Emitter follower configuration is very useful for impedance matching applications because of the very high input impedance, in the region of hundreds of thousands of Ohms, and it has relatively low output impedance.

Common Collector Amplifier Circuit:

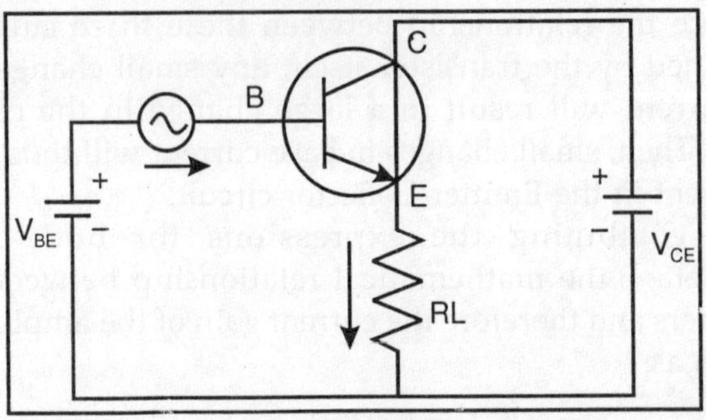

The Common Emitter configuration has a current gain equal to the β value of the transistor itself. In the common collector configuration the load resistance is situated in series with the emitter so its current is equal to that of the emitter current.

As the emitter current is the combination of the collector AND base currents combined, the load resistance in this type of amplifier configuration also has both the collector current and the input current of the base flowing through it. Then the current gain of the circuit is given as:

$$I_E = I_C + I_B$$

$$A_i = \frac{I_E}{I_B} = \frac{I_C + I_B}{I_B}$$

$$A_i = \frac{I_C}{I_B} + 1$$

$$\therefore A_i = \beta + 1$$

This type of bipolar transistor configuration is a non-inverting amplifier circuit in that the signal voltages of Vin and Vout are "In-Phase". It has a voltage gain that is always less than "1" (unity). The load resistance of the common collector amplifier configuration receives both the base and collector currents giving a large current gain (as with the Common Emitter configuration) therefore, providing good current amplification with very little voltage gain.

FUNDAMENTALS OF BJT OPERATION

A bipolar junction transistor consists of two back-to-back p-n junctions, who share a thin common region with width, w_B. Contacts are made to all three regions, the two outer regions called the emitter and collector and the middle region called the base. The structure of an npn bipolar transistor is shown in Figure below (a). The device is called "bipolar" since its operation involves both types of mobile carriers, electrons and holes.

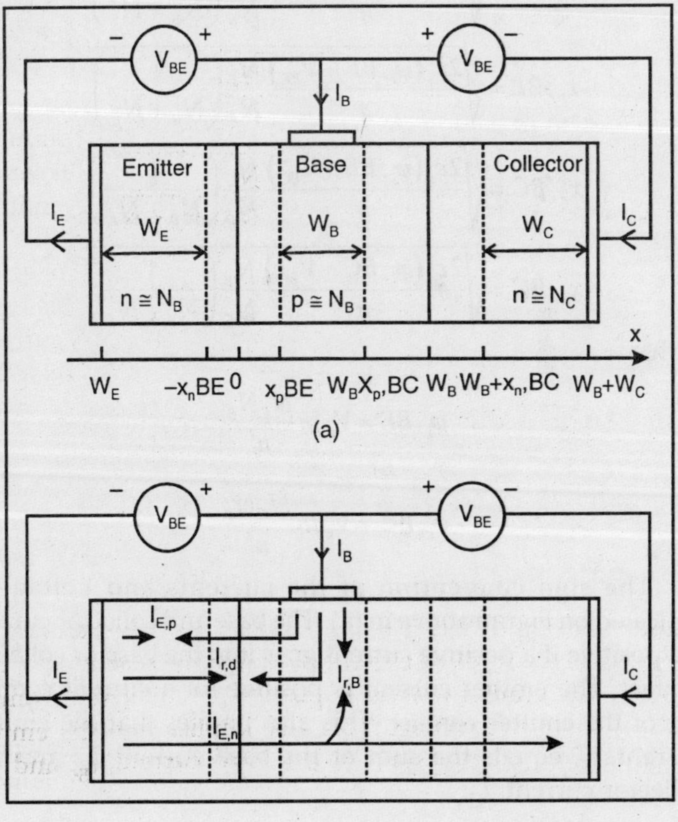

(a) Structure and sign convention of a npn bipolar junction transistor. (b) Electron and hole flow under forward active bias, $V_{BE} > 0$ and $V_{BC} = 0$. Since the device consists of two back-to-back diodes, there are depletion regions between the quasi-neutral regions. The width of the quasi neutral regions in the emitter, base and collector are indicated with the symbols w_E', w_B' and w_C' and are calculated from

$$w_E' = w_E - x_n, BE$$

$$w_B' = w_B - x_p, BE^{-x} p, BC$$

$$w_C' = w_C - x_n, BC$$

where the depletion region widths are given by:

$$x_n, BE = \sqrt{\frac{2\varepsilon_s(\varphi_i, BE - V_{BE})}{q}\frac{N_B}{N_E}\left(\frac{1}{N_B + N_E}\right)}$$

$$x_p, BE = \sqrt{\frac{2\varepsilon_s(\varphi_i, BE - V_{BE})}{q}\frac{N_E}{N_B}\left(\frac{1}{N_B + N_E}\right)}$$

$$x_p, BC = \sqrt{\frac{2\varepsilon_s(\varphi_i, BE - V_{BC})}{q}\frac{N_C}{N_B}\left(\frac{1}{N_B + N_E}\right)}$$

$$x_n, BC = \sqrt{\frac{2\varepsilon_s(\varphi_i, BC - V_{BC})}{q}\frac{N_B}{N_C}\left(\frac{1}{N_B + N_E}\right)}$$

with,

$$\varphi_i, BE = V_t \, In \frac{N_E N_B}{n_i^2}$$

$$\varphi_i, BC = V_t \, In \frac{N_C N_B}{n_i^2}$$

The sign convention of the currents and voltage is indicated on Figure above in (a). The base and collector current are positive if a positive current goes into the base or collector contact. The emitter current is positive for a current coming out of the emitter contact. This also implies that the emitter current, I_E, equals the sum of the base current, I_B, and the collector current, I_C:

$$I_E = I_C + I_B$$

The base-emitter voltage and the base-collector voltage are positive if a positive voltage is applied to the base contact relative to the emitter and collector respectively.

The operation of the device is illustrated with Figure above (b). We consider here only the forward active bias mode of operation, obtained by forward biasing the base-emitter junction and reverse biasing the base-collector junction.

To simplify the discussion further, we also set $V_{CE} = 0$. The corresponding energy band diagram is shown in Figure below. Electrons diffuse from the emitter into the base and holes diffuse from the base into the emitter.

This carrier diffusion is identical to that in a p-n junction. However, what is different is that the electrons can diffuse as minority carriers through the quasi-neutral base. Once the electrons arrive at the base-collector depletion region, they are swept through the depletion layer due to the electric field.

These electrons contribute to the collector current. In addition, there are two more currents, the base recombination current, indicated on Figure below by the vertical arrow, and the base-emitter depletion layer recombination current, $I_{r,d}$, (not shown).

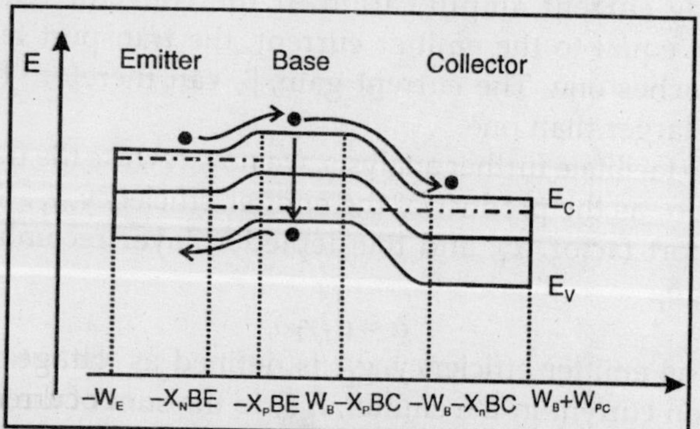

Energy band diagram of a bipolar transistor biased in the forward active mode.

The total emitter current is the sum of the electron diffusion current, $I_{E,n}$, the hole diffusion current, $I_{E,p}$ and the base-emitter depletion layer recombination current, $I_{r,d}$.

$$I_E = I_{E,n} + I_{E,P} + I_{r,d}$$

The total collector current is the electron diffusion current, $I_{E,n}$, minus the base recombination current, $I_{r,B}$.

$$I_C = I_{E,n} - I_{r,B}$$

The base current is the sum of the hole diffusion current, $I_{E,p}$, the base recombination current, $I_{r,B}$ and the base-emitter depletion layer recombination current, $I_{r,d}$.

$$I_B = I_{E,P} + I_{r,B} + I_{r,d}$$

The transport factor, α, is defined as the ratio of the collector and emitter current

$$\alpha = \frac{I_C}{I_E}$$

Using Kirchoff's current law and the sign convention shown in Figure above in (a), we find that the base current equals the difference between the emitter and collector current. The current gain, β, is defined as the ratio of the collector and base current and equals:

$$\beta = \frac{I_C}{I_E} = \frac{\alpha}{1-\alpha}$$

This explains how a bipolar junction transistor can provide current amplification. If the collector current is almost equal to the emitter current, the transport factor, a, approaches one. The current gain, β, can therefore become much larger than one.

To facilitate further analysis, we now rewrite the transport factor, α, as the product of the emitter efficiency, γ_E, the base transport factor, α_T, and the depletion layer recombination factor, δ_r.

$$\alpha = \alpha_T \gamma_E \delta_r$$

The emitter efficiency, g_E, is defined as the ratio of the electron current in the emitter, $I_{E,n}$, to the sum of the electron and hole current diffusing across the base-emitter junction, $I_{E,n} + I_{E,p}$.

$$\gamma_E = \frac{I_{E.n}}{I_{E,N} + I_{E,p}}$$

The base transport factor, a_T, equals the ratio of the current due to electrons injected in the collector, to the current due to electrons injected in the base.

$$\alpha_T = \frac{I_{E.n} - I_{r,B}}{I_E, n}$$

Recombination in the depletion-region of the base-emitter junction further reduces the current gain, as it increases the emitter current without increasing the collector current. The depletion layer recombination factor, d_r, equals the ratio of the current due to electron and hole diffusion across the base-emitter junction to the total emitter current:

$$\delta_r = \frac{I_E - I_{r,d}}{I_E}$$

Example: A bipolar transistor with an emitter current of 1 mA has an emitter efficiency of 0.99, a base transport factor of 0.995 and a depletion layer recombination factor of 0.998. Calculate the base current, the collector current, the transport factor and the current gain of the transistor.

Solution:

The transport factor and current gain are:

$$\alpha = \gamma E \alpha T \delta_\gamma = 0.99 \times 0.995 \times 0.998 = 0.9983$$

and

$$\beta = \frac{\alpha}{1-\alpha} = 58.1$$

The collector current then equals

$$I_C = \alpha I_E = 0.983 mA$$

And the base current is obtained from:

$$I_B = I_E - I_C = 17\,\mu A$$

AMPLIFICATIONS

Amplification Classes (III)

When communicating with another ham the main objective is to emit a signal displaying an excellent modulation that please to your correspondent's ears. Single-sideband transmitters generate the carrier and modulation at a low frequency and translate it up with a mixer. Linear amplifiers are used to increase the power level up to say 100W.

If we have to amplify this signal to higher values because the station if far away, we want that our amplified signal be strictly identical in shape to the original, only its intensity must have increased. We call this characteristics, the linearity.

Linearity

Several factors determine the linearity of an HF so-called "linear" amplifier:

- Changes in device power gain (ratio of output to drive power) over the range from zero to full output
- The collector/emitter (BJT transistor) or drain/source (MOSFET transistor) peak RF voltage excursion
- The regulation (stiffness) of the collector (+Vcc) or drain (+Vdd) supply voltage
- The standing (idle) current and the onset of saturation in the RF transformers.

The limiting factor is the constancy of power gain over the entire power-output excursion. To visualise this, one can inspect the curve of output vs. drive power in the transistor data sheet. Generally, MOSFETs will exhibit superior linearity as compared to bipolar junction transistors (BJTs). MOSFETs such as the Motorola MRF150 began to displace BJTs (e.g. MRF448, 2SC2652) in the late 1990's.

The higher the voltage excursion, the longer the linear portion of the output/drive power curve. For this reason, amplifiers powered between 40-50V DC exhibit considerably better linearity, and thus lower IMD, than 13.8V units. Typical 3rd-order IMD (IMD3) values relative to PEP are –32 dB for 50V and –24 dB for 13.8V. Additionally, it is more difficult to design a power supply for a 13.8 V amplifier due to high current requirements (e.g. typically 80A peak for a 500W PEP amplifier like Ameritron ALS-500MX). So usually a tube amplifier is more linear than any solid-state model (typically IMD3 < –40 dB relative to PEP).

The collector-to-collector (or drain-to-drain) load impedance must be maintained as close to resistive as possible.

This in turn requires optimising the wideband output transformer(s), combiner (if used), and low-pass filter passband VSWRs to as low values as possible.

The load presented to the low-pass filter output must also be as close as possible to 50 ohms resistive. The amplification class defines the functioning mode of the electronic circuit. It has been divided in several classes: A, B, AB, AB1, AB2, C, D, E, G, H.

Class A is the most appreciated by fans of high fidelity audio and amateur radio because this is the mode in which the signal yield the best performances, in both musicality (for audio applications) and output. However this class has some drawbacks and therefore hams prefer using power devices in Class AB, so many classes that we are going to review.

Class A

To work in Class A, typically the signal to amplify has a working point (quiescent point) that normally lies in the middle of the active region of the load line, as displayed at left. At rest, without input signal, the cathode current flows contineously in the device (tube or transistor), consuming at lot of power, proportional to $V_{ce}I_c$. One says that this mode offers a conducting angle of 360°.

To avoid distorsions (saturation or blocking) the peak values for the output current and voltage must be always smaller than the working point voltage and current. The maximum signal power of this device is limited by the relationships:

$$P_{\max} = \frac{U_{CE} I_C}{2} P_{dc} = U_b I_c$$

In Class A the output signal is directly proportional to the input; hence the name of linear amplifier. This mode is ideal if you are looking for high quality signals but it is paid by the consumption of much energy.

Its efficiency is at best 50%. That means that more than a half of the power is lost in heat, and indeed these amplifiers heat much and request an oversized radiator to dissipate the lost power.

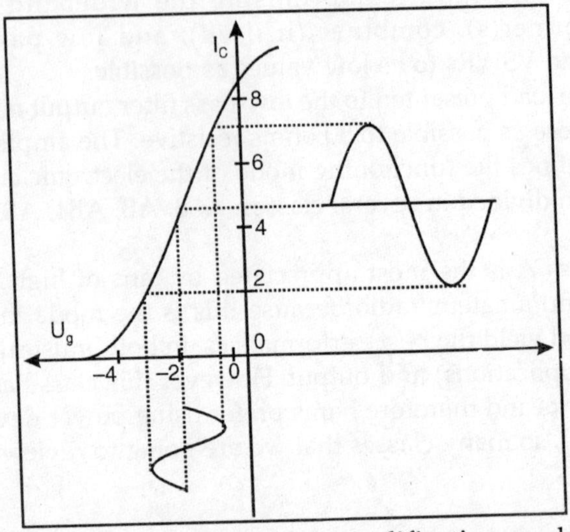

Class A is the widely used amplification mode in low frequencies and in the first HF stages of radio equipments. And as you probably know if you have a Hi-Fi system, in audio it is the privileged amplification class for the purists.

Class B

Class A or Class B amplifiers are used to increase the power level of weak AM or SSB signals. But Class A amplification yield a low output due to much power dissipation.

The best should be to suppress the polarization current in quiescent mode. This is achieved in Class B amplifiers that are biased at cutoff (blocking point) so that the collector current flows for 180° of the input cycle. In other words, as soon as there is an input signal, the device unlocks during half a cycle and conducts.

If we suspect that this tube or transistor will yield a better output as it does no more consume power at rest (quiescent mode), the output signal does only return one alternation of two and we will guess that this mode is not satisfying. Then, as the device will only work in the presence of an input signal, for weak inputs it will not work in the linear region of its characteristics and thus it will not be linear.

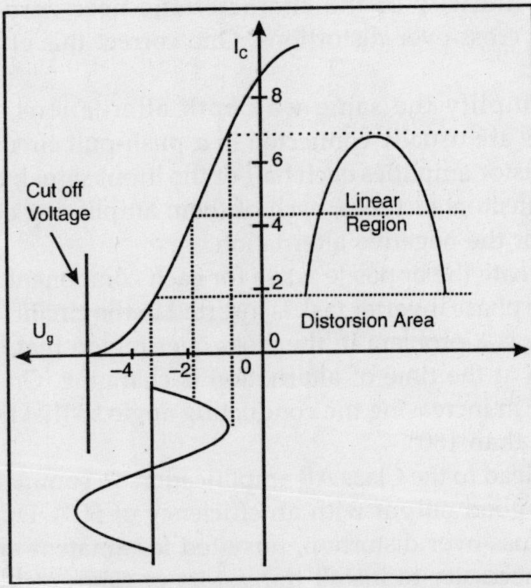

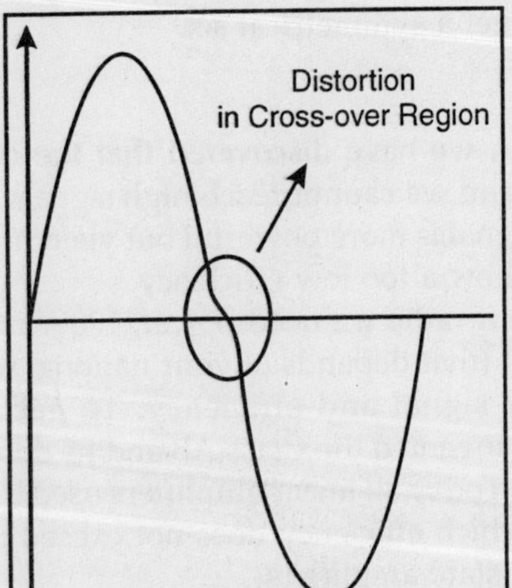

At left when a Class B amplifier works close to cut off, its transistors or tubes can only amplify one half of the signal. Therefore class B amplifiers will always consist of two transistors or tubes (push-pull circuit), where each component will supply one half wave of the output signal. At right the control characteristics of two transistors of a Class B amplifier.

The non-linearity of the characteristic near zero-current produces cross-over distortions. One correct this effect with Class AB.

To amplify the same way both alternations, Class B amplifiers are usually connected in a push-pull circuit where one transistor amplifies each half of the input signal; there are thus 2 transitors or tubes, each of them amplifying either the positive or the negative alternation.

To create the opposite wave for each component, a phase splitter or phase inverter task is inserted in the circuit. But here also there is a problem in the cross-over region that creates a distortion at the time of alternation is changing. One correct this effect in increasing the conducting angle so that the phase is greater than 180°.

This lead to the Class AB amplification. In summary, Class B yield a good output with an efficiency of 60%. Drawbacks are the cross-over distortion, unsuited for amateur emissions and the necessity to install transistors or tubes controlled in anti-phase to get a symmetrical set.

Class AB (IV)

In Class A we have discovered that the ouput power is quite limited and we cannot reach high signals while in Class B the ouput signal is more powerful but yield no linearity, and both classes show a too low efficiency.

As amateur radio we need power, 100W in input and up to 2 kW ouput (that depends of your national regulation), we need a linear signal and efficiency. To get these results, electronicians invented the Class AB and its derivated Classes AB1 and AB2. Today all linear amplifiers used by hams belong to Class AB which efficiency does not exceed 50%, there are tubes or solid-state amplifiers.

These amplifiers use power devices that are are slightly polarized so that they work in the linear region of the load line for weak signals (Class A) and in Class B for stronger signals. In this way we avoid cross-over distortions as the working point of transistors or tubes can be placed a bit higher than the cut off at the lower end of the active region.

This polarization is less than in Class A, allowing the transistor or tube to emit very few current when there is no signal in input. But as they are somewhat, the linearity is modified. Therefore in all amateurs amplifiers the output circuit is completed with a filtering function to cut all harmonics produced in this process.

Class AB is subdivided in Class AB1 and AB2 depending the power stage characteristics Class AB1 concerns a power stage that does not absorb current from the previous stage; this is usually the case with vacuum tubes without grid current and the Field Effect Transistors, the famous FETs. This stage yield usually a high impedance Class AB2 concerns a power stage that absorbs some current from the previous stage; some vaccuum tubes are used in this manner and all the classic bipolar transistors.

In the field it is easy to recognize a ham using a Class AB amplifier so-called linear. When speaking some of his sentences are grinded and works lack in the conversation. This effect occurs when the polarization is too low, with for consequence to transmit only the peaks of modulation; the current is transmitted during less than half of a wave of the output signal. To solve this problem this ham has to increase the quiescent current.

Chapter 5

Special Devices

PHOTODIODES

INTRODUCTION

Photodiodes are semiconductor light sensors that generate a current or voltage when the P-N junction in the semiconductor is illuminated by light. The term photodiode can be broadly defined to include even solar batteries, but it usually refers to sensors used to detect the intensity of light. Photodiodes can be classified by function and construction as follows:

PHOTODIODE TYPE

- PN photodiode
- PIN photodiode
- Schottky type photodiode
- APD (Avalanche photodiode)

All of these types provide the following features and are widely used for the detection of the intensity, position, colour and presence of light.

FEATURES OF PHOTODIODE

- Excellent linearity with respect to incident light
- Low noise
- Wide spectral response
- Mechanically rugged
- Compact and lightweight
- Long life

PRINCIPLE OF OPERATION

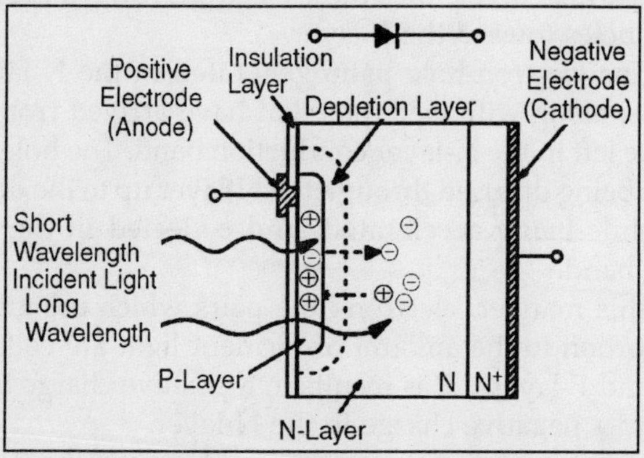

Fig. Below Shows a Cross Section of a Photodiode.

The P-layer material at the active surface and the N material at the substrate form a PN junction which operates as a photoelectric converter. The usual P-layer for a Si photodiode is formed by selective diffusion of boron, to a thickness of approximately 1 μm or less and the neutral region at the junction between the P- and N-layers is known as the depletion layer. By controlling the thickness of the outer P-layer, substrate N-layer and bottom N^+-layer as well as the doping concentration, the spectral response and frequency response can be controlled. When light strikes a photodiode, the electron within the crystal structure becomes stimulated. If the light energy is greater than the band gap energy Eg, the electrons are pulled up into the conduction band, leaving holes in their place in the valence band as shown below:

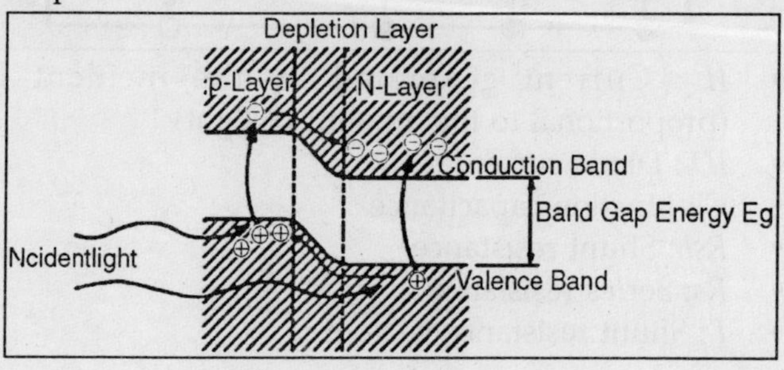

These electron-hole pairs occur throughout the P-layer, depletion layer and N-layer materials. In the depletion layer the electric field accelerates these electrons toward the N-layer and the holes toward the P-layer.

Of the electron-hole pairs generated in the N-layer, the electrons, along with electrons that have arrived from the P-layer, are left in the N-layer conduction band. The holes at this time are being diffused through the N-layer up to the depletion layer while being accelerated, and collected in the P-layer valence band.

In this manner, electron-hole pairs which are generated in proportion to the amount of incident light are collected in the N- and P-layers. This results in a positive charge in the P-layer and a negative charge in the N-layer.

If an external circuit is connected between the P- and N-layers, electrons will flow away from the N-layer, and holes will flow away from the P-layer toward the opposite respective electrodes. These electrons and holes generating a current flow in a semiconductor are called the carriers.

SI PHOTODIODE

An equivalent circuit of a photodiode is shown in Figure.

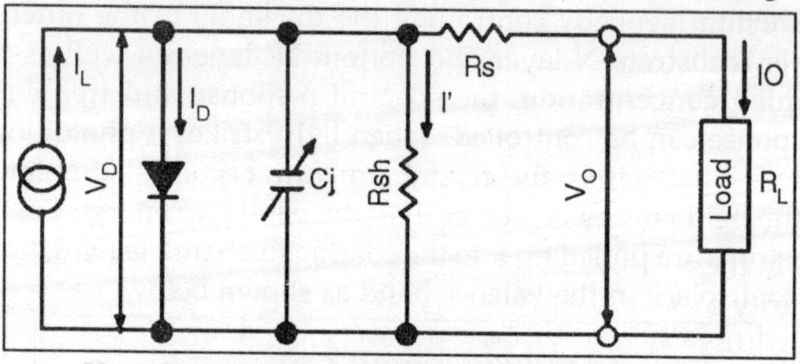

- *IL* : Current generated by the incident light (proportional to the amount of light)
- *ID* : Diode current
- *Cj*: Junction capacitance
- *Rsh*: Shunt resistance
- *Rs*: Series resistance
- *I'*: Shunt resistance current

- *VD* : Voltage across the diode
- *Io*: Output current
- *Vo*: Output voltage

Using the above equivalent circuit, the output current Io is given as follows:

$$Io = IL - ID - I' = IL - Is\left(\exp\frac{eVD}{kT} - 1\right) - I' ......(2-1)$$

- *Is*: Photodiode reverse saturation current
- *e*: Electron charge
- *k*: Boltzmann's constant
- *T*: Absolute temperature of the photodiode

The open circuit voltage Voc is the output voltage when Io equals 0.

Thus Voc becomes:

$$Voc = \frac{kT}{e} In\left(\frac{IL - I}{Is} + 1\right) ......(2-2)$$

If I' is negligible, since Is increases exponentially with respect to ambient temperature, Voc is inversely proportional to the ambient temperature and proportional to the log of IL. However, this relationship does not hold for very low light levels.The short circuit current Isc is the output current when the load resistance RL equals 0 and Vo equals 0, yielding:

$$Isc = IL - Is\left(\exp\frac{(Isc \cdot Rs)}{kT} - 1\right) - \frac{Isc \cdot Rs}{Rsh} ....(2-3)$$

In the above relationship, the 2nd and 3rd terms limit the linearity of Isc. However, since Rs is several ohms and Rsh is 10^7 to 10^{11} ohms, these terms become negligible over quite a wide range.

Current vs. Voltage Characteristic

When a voltage is applied to a photodiode in the dark state, the current vs. voltage characteristic observed is similar to the curve of a conventional rectifier diode as shown in Figure.

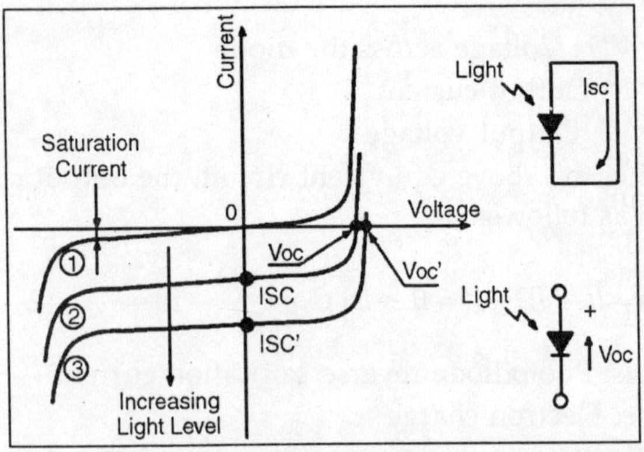

However, when light strikes the photodiode, the curve at (1) shifts to (2) and, increasing the amount of incident light shifts this characteristic curve still further to position (3) in parallel, according to the incident light intensity. As for the characteristics of (2) and (3), if the photodiode terminals are shorted, a photocurrent Isc or Isc proportional to the light intensity will flow in the direction from the anode to the cathode.

If the circuit is open, an open circuit voltage Voc or Voc′ will be generated with the positive polarity at the anode.The short circuit current Isc is extremely linear with respect to the incident light level. When the incident light is within a range of 10^{-12} to 10^{-2} W, the achievable range of linearity is higher than 9 orders of magnitude, depending on the type of photodiode and its operating circuit. The lower limit of this linearity is determined by the NEP, while the upper limit depends on the load resistance and reverse bias voltage, and is given by the following equation:

$$Psat = \frac{V_{Bi} + V_R}{(Rs + R_L) \cdot S\lambda} \text{.......}(2-4)$$

- *Psat:* Input energy (W) at upper limit of linearity Psat £10 mW
- *VBτ:* Contact voltage (V) <0.2 to 0.3 V>
- *VR*: Reverse voltage (V)
- *RL* : Load resistance (Ω)
- *Sλ*: Photo sensitivity at wavelength λ (A/W)

- *Rs*: Photodiode series resistance (several Ω)

When laser light is condensed on a small spot, however, the actual series resistance element increases, and linearity deteriorates. Voc varies logarithmically with respect to a change of the light level and is greatly affected by variations in temperature, making it unsuitable for light intensity measurements. Figure below shows the result of plotting Isc and Voc as a function of incident light illuminance.

(a) Short circuit current

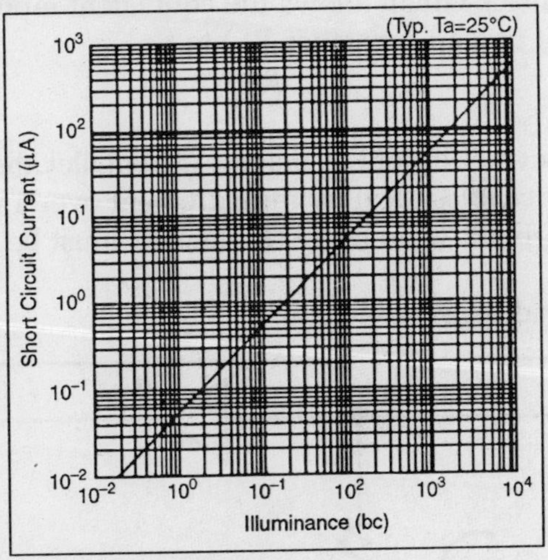

(b) Open circuit voltage

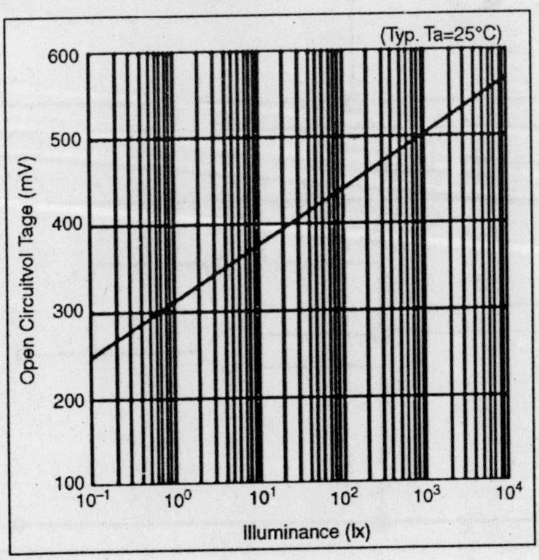

Figures (a) and (b) shown below are the methods of measuring light by measuring the photocurrent IL or Isc. In the circuit shown at (a), the voltage (Io × RL) is amplified by an amplifier with gain G, although the circuit does have limitations on its linearity according to equation (2-4). This condition is shown in Figure as shown below. Figure below (b) is a circuit using an operational amplifier. If we set the open loop gain of the operational amplifier as A, the characteristics of the feedback circuit allows the equivalent input resistance (equivalent to load resistance RL) to be,

$$\frac{Rf}{A}$$

which is several orders of magnitude smaller than Rf. Thus this circuit enables ideal Isc measurement over a wide range. For measuring a wide range, RL and Rf must be adjusted as needed.

(a) Load resistance circuit

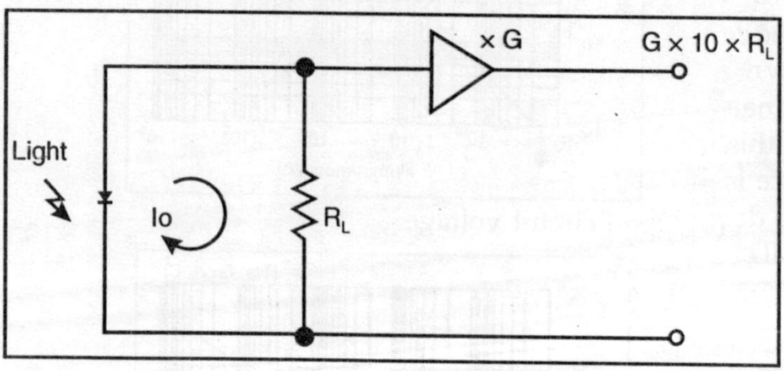

(b) Op-amp circuit

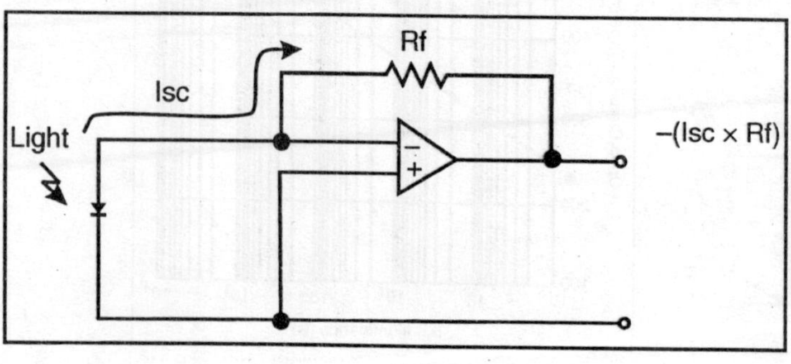

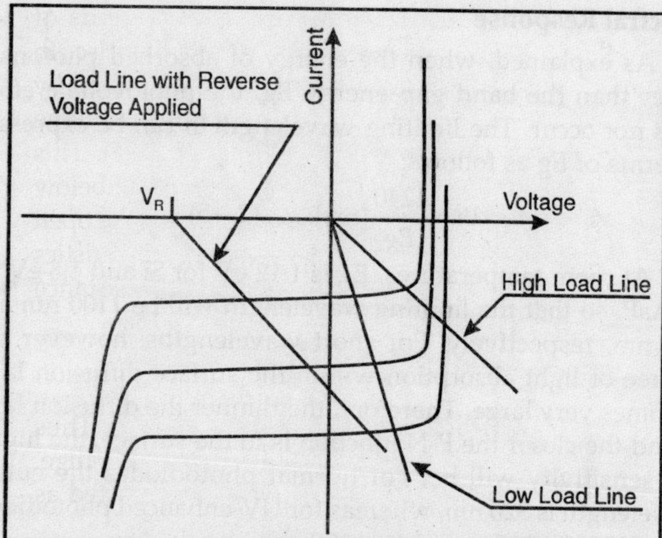

Fig Current vs. Voltage Characteristic and Load Line

If the zero region of Figure above (1) is magnified, we see, as shown in Figure below, that the dark current ID is approximately linear in a voltage range of about ±10 mV. The slope in this region indicates the shunt resistance Rsh and this resistance is the cause of the thermal noise current described later. In data sheets, values of Rsh are given using a dark current ID measured with –10 mV applied.

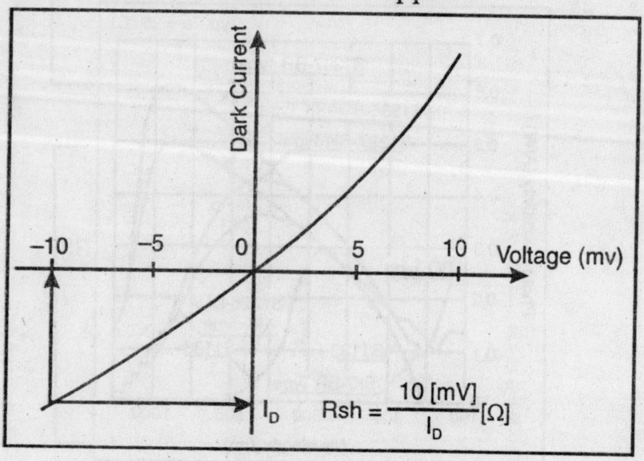

Spectral Response

As explained, when the energy of absorbed photons is lower than the band gap energy Eg, the photovoltaic effect does not occur. The limiting wavelength lh can be expressed in terms of Eg as follows:

$$\lambda h = \frac{1240}{Eg}[nm]........(2-5)$$

At room temperatures, Eg is 1.12 eV for Si and 1.8 eV for GaAsP, so that the limiting wavelength will be 1100 nm and 700 nm, respectively. For short wavelengths, however, the degree of light absorption within the surface diffusion layer becomes very large. Therefore, the thinner the diffusion layer is and the closer the P-N junction is to the surface, the higher the sensitivity will be. For normal photodiodes the cut-off wavelength is 320 nm, whereas for UV-enhanced photodiodes (e.g. S1226/S1336 series) it is 190 nm.

The cut-off wavelength is determined by the intrinsic material properties of the photodiode, but it is also affected by the spectral transmittance of the window material. For borosilicate glass and plastic resin coating, wavelengths below approximately 300 nm are absorbed. If these materials are used as the window, the short wavelength sensitivity will be lost. For wavelengths below 300 nm, photodiodes with quartz windows are used.

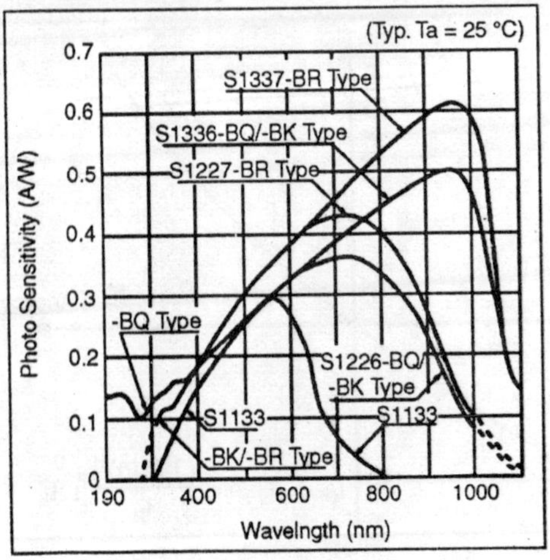

For measurements limited to the visible light region, a visual-compensation filter is used as the light-receiving window. Figure above shows the spectral response characteristics for various photodiode types.

The BQ type shown uses a quartz window, the BK type a borosilicate glass window and the BR type a resin-coated window. S1133 is a visible photodiode with a visual-compensated filter.

Noise Characteristic

Like other types of light sensors, the lower limits of light detection for photodiodes are determined by the noise characteristics of the device.

The photodiode noise in is the sum of the thermal noise (or Johnson noise) *ij* of a resistor which approximates the shunt resistance and the shot noise isD and isL resulting from the dark current and the photocurrent.

$$in = \sqrt{ij^2 + isD^2 + isL^2}\,[A] \text{..........} (2-6)$$

ij is viewed as the thermal noise of Rsh and is given as follows:

$$ij = \sqrt{\frac{4kTB}{Rsh}}\,[A] \text{..........} (2-7)$$

K: Boltzmann's Constant

T: Absolute temperature of the element

B: Noise bandwidth

When a bias voltage is applied as in Figure 3-1, there is always a dark current. The shot noise *isD* originating from the dark current is given by,

$$isD = \sqrt{2qIDB}\,[A] \text{..........} (2-8)$$

q: Electron charge

ID: Dark current

B: Noise bandwidth

With the application of incident light, a photocurrent IL exists so *isL* is given by,

$$isL = \sqrt{2qILB}\,[A] \text{..........} (2-9)$$

If IL >> 0.026/Rsh or IL >> ID, the shot noise current of

equation (2–9) becomes predominant instead of the noise factor of eqaution (2–7) or (2–8).

The amplitudes of these noise sources are each proportional to the square root of the measured bandwidth B so that they are expressed in units of $A/Hz^{1/2}$.

The lower limit of light detection for a photodiode is usually expressed as the intensity of incident light required to generate a current equal to the noise current as expressed in equation (2–7) or (2–8). Essentially this is the noise equivalent power (NEP).

$$NEP = \frac{in}{S}\left[W/Hz^{1/2}\right] \text{.........}(2-10)$$

in: Noise current ($A/Hz^{1/2}$)
S: Photo sensitivity (A/W)

Figure below shows the relationship between NEP and shunt resistance, from which a photodiode is agreement with the theoretical relationship.

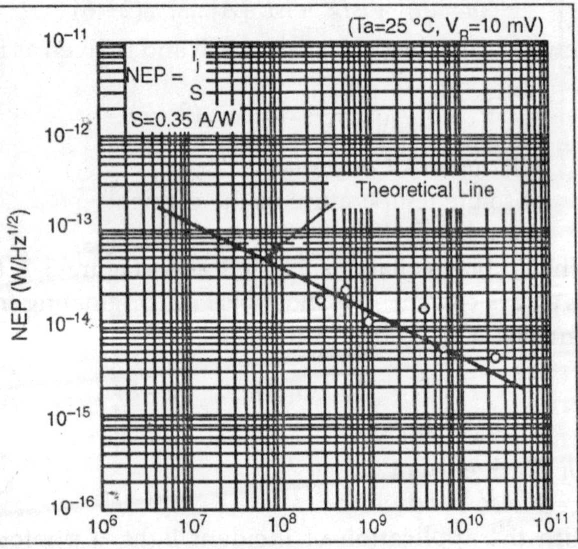

Spatial Response Uniformity

This is the measure of the variation in sensitivity with the position of the active area. Photodiodes offer excellent uniformity, usually less than 1 %. This uniformity is measured with light from a laser diode (680 nm) condensed to a small spot from several microns to several dozen microns in diameter.

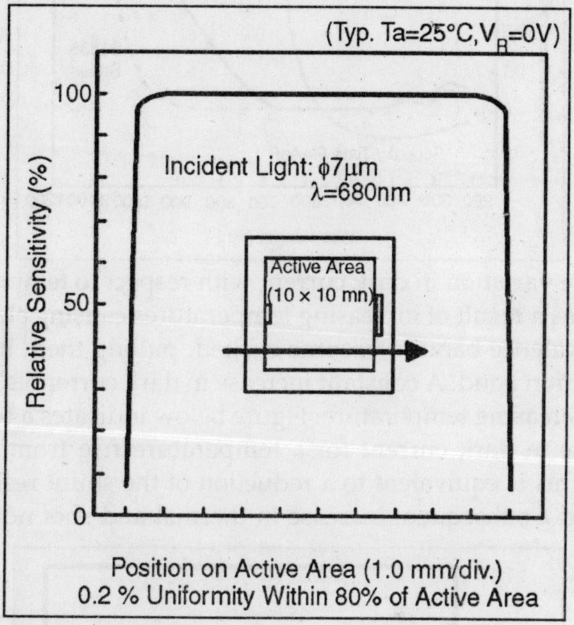

Temperature Characteristics

Ambient temperature variations greatly affect photodiode sensitivity and dark current. The cause of this is variation in the light absorption coefficient which is temperature related.

For long wavelengths, sensitivity increases with increasing temperature and this increase become prominent at wavelengths longer than the peak wavelength.

For short wavelengths, it decreases. Since ultraviolet enhanced photodiodes are designed to have low absorption in the short wavelength region, the temperature coefficient is extremely small at wavelengths shorter than the peak wavelength. Figure below shows examples of temperature coefficients of photodiodes sensitivity for a variety of photodiodes types.

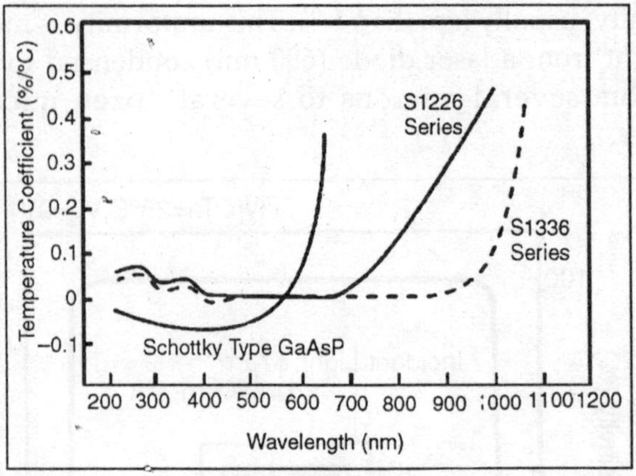

The variation in dark current with respect to temperature occurs as a result of increasing temperatures causing electrons in the valence band to become excited, pulling them into the conduction band. A constant increase in dark current is shown with increasing temperature. Figure below indicates a twofold increase in dark current for a temperature rise from 5°C to 10°C. This is equivalent to a reduction of the shunt resistance Rsh and a subsequent increase in thermal and shot noise.

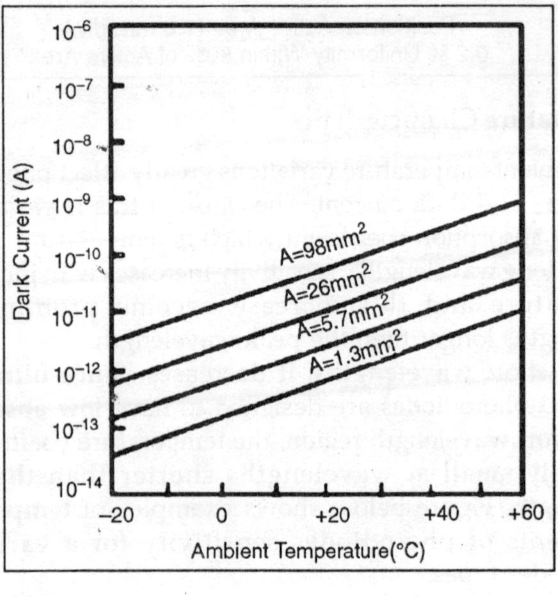

Figure below shows an example of the temperature characteristics of open-circuit voltage Vop, indicating linearity with respect to temperature change.

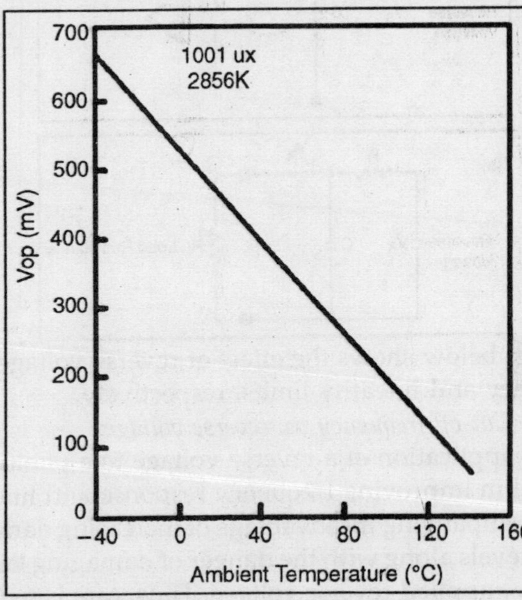

SI PIN PHOTODIODE

Reverse Voltage

Because photodiodes generate a power due to the photovoltaic effect, they can operate without the need for an external power source.

However, frequency response and linearity can be improved by using an external reverse voltage VR. It should be borne in mind that the signal current flowing in a photodiode circuit is determined by the number of photovoltaically generated electron-hole pairs and that the application of a reverse voltage does not affect the signal current nor impair the photoelectric conversion linearity.

Figure below shows examples of reverse voltage connection.

Reverse voltage connection:

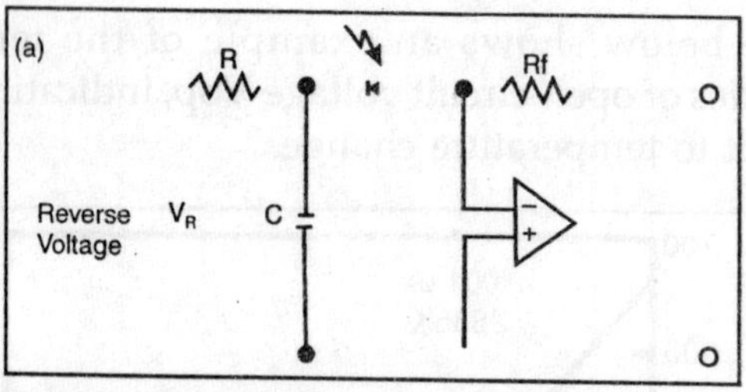

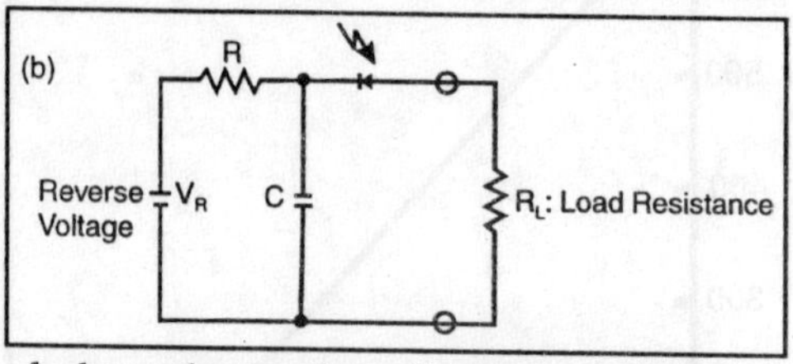

Figures below shows the effect of reverse voltage on cut-off frequency and linearity limits, respectively.

Figure Cut-off frequency vs. reverse voltage:

While application of a reverse voltage to a photodiode is very useful in improving frequency response and linearity, it has the accompanying disadvantage of increasing dark current and noise levels along with the danger of damaging the device by excessive applied reverse voltage. Thus, care is required to maintain the reverse voltage within the maximum ratings and to ensure that the cathode is maintained at a positive potential with respect to the anode.

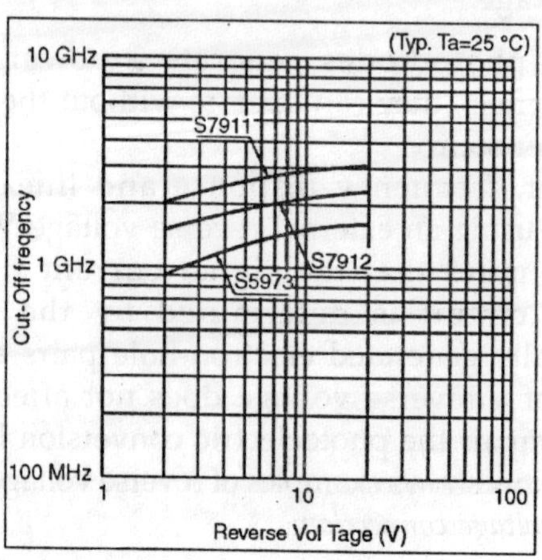

For use in applications such as optical communications and remote control which require high response speed, the PIN photodiode provides not only good response speed but excellent dark current and voltage resistance characteristics with reverse voltage applied.

Note that the reverse voltages listed in data sheets are recommended values and each PIN photodiode is designed to provide optimum perform-ance at the recommended reverse voltage.

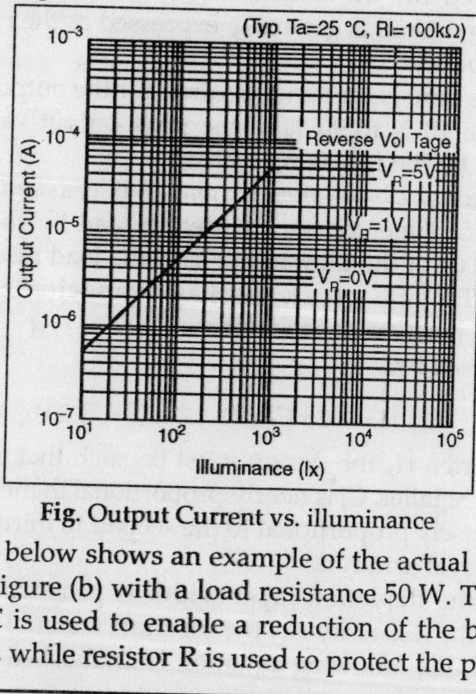

Fig. Output Current vs. illuminance

Figure below shows an example of the actual connection shown in Figure (b) with a load resistance 50 W. The ceramic capacitor C is used to enable a reduction of the bias supply impedance, while resistor R is used to protect the photodiode.

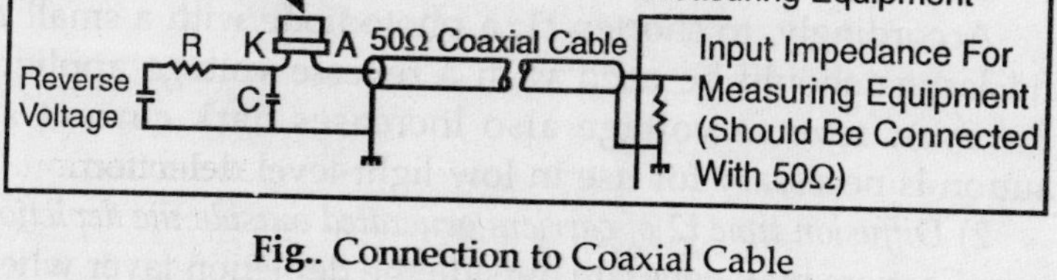

Fig.. Connection to Coaxial Cable

The resistor value is selected such that the voltage drop

caused by the maximum photocurrent is sufficiently smaller than the reverse voltage.

The photodiode and capacitor leads, coaxial cable and other wire carrying high-speed pulses should be kept as short as possible.

Response Speed and Frequency Response

The response speed of a photodiode is a measure of the time required for the accumulated charge to become an external current and is generally expressed as the rise time or cut-off frequency.

The rise time is the time required for the output signal to change from 10 % to 90 % of the peak output value and is determined by the following factors:

1) *Terminal capacitance Ct and time constant t1 of load resistance RL:* Time constant t1 determined by the terminal capacitance Ct of the photodiode and the load resistance RL.

Ct is the sum of the package capacitance and the photodiode junction capacitance.

t1 is given by,

$$t1 = 2.2 \times Ct \times RL ....(3-1)$$

To shorten t1, the design must be such that either Ct or RL is made smaller. Cj is nearly proportional to the active area A and inversely proportional to the second to third root of the depletion layer width d.

Since the depletion layer width is proportional to the product of the resistivity r of the substrate material and reverse voltage VR, the following equation is established as:

$$Cj \propto A\{(VR + 0.5) \times \rho\}^{-1/2 to -1/3} .....(3-2)$$

Accordingly, to shorten t1, a photodiode with a small A and large r should be used with a reverse voltage applied. However, reverse voltage also increases dark current so caution is necessary for use in low-light-level detection.

2) *Diffusion time t2 of carriers generated outside the depletion layer*: Carriers may generate outside the depletion layer when incident light misses the P-N junction and is absorbed by the

surrounding area of the photodiode chip and the substrate section which is below the depletion area. The time t2 required for these carriers to diffuse may sometimes be greater than several microseconds.

3) *Carrier transit time t3 in the depletion layer*: The transit speed vd at which the carriers travel in the depletion layer is expressed using the traveling rate μ and the electric field E developed in the depletion layer, as in vd = μE.

If we let the depletion layer width be d and the applied voltage be VR, the average electric field E=VR/d, and thus t3 can be approximated as follows:

$$t3 = d / vd = d^2 / (\mu VR) \text{........} (3-3)$$

To achieve a fast response time for t3, the moving distance of carriers should be short and the reverse voltage larger.

The above three factors determine the rise time tr of a photodiode and rise time tr is approximated by the following equation:

$$tr = \sqrt{t1^2 + t2^2 + t3^2} \text{......} (3-4)$$

PIN photodiodes and avalanche photodiodes are designed such that less carriers are generated outside the depletion layer, Ct is small and the carrier transit time in the depletion layer is short. Therefore, these types are ideally suited for high-speed light detection.

The cut-off frequency fc is the frequency at which the photodiode output decreases by 3 dB from the output at 100 kHz when the photodiode receives sinewave-modulated light from a laser diode.

The rise time tr roughly approximates this fc in the formula:

$$tr = \frac{0.35}{fc} \text{......} (3-5)$$

Figures (a), (b) and (c) below show examples of the response waveform and frequency response character-istics for typical photodiodes.

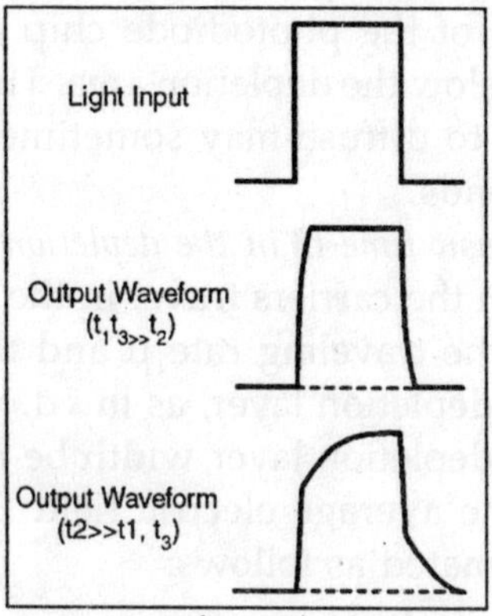

Fig. (a) Photodiode Response Waveform Example

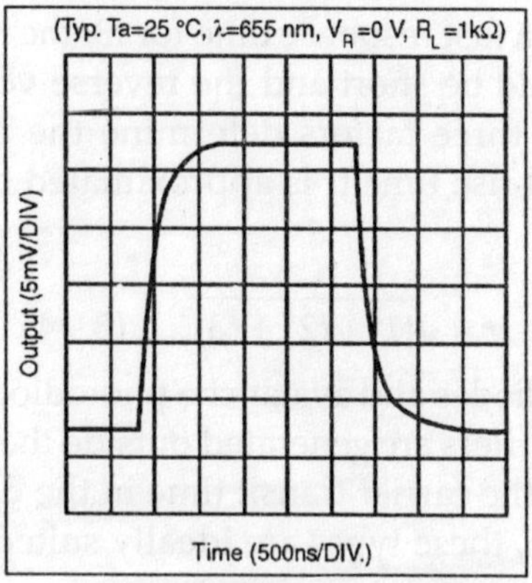

Fig. (b) Response Waveform (S2386-18K)

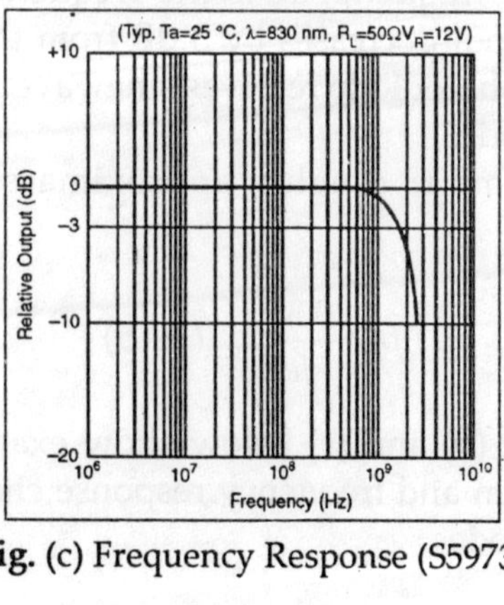

Fig. (c) Frequency Response (S5973)

SI PHOTODIODE WITH PREAMP

Feedback Circuit

Figure below shows a basic circuit connection of an operational amplifier and photodiode.

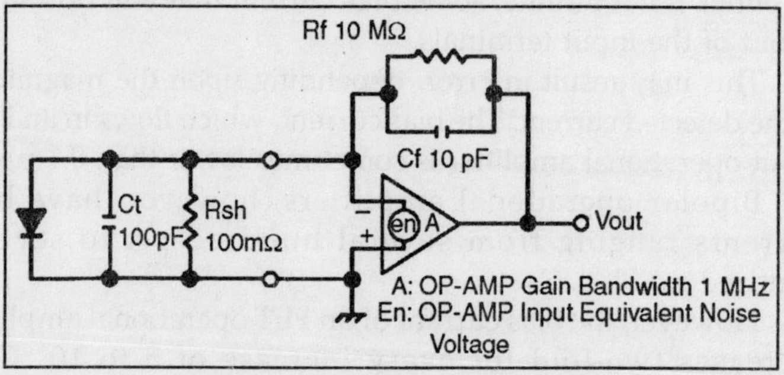

Fig. Basic Photodiode Connection

The output voltage Vout from DC through the low-frequency region is 180 degrees out of phase with the input current Isc.

The feedback resistance Rf is determined by Isc and the required output voltage Vout. If, however, Rf is made greater than the photodiode internal resistance Rsh, the operational amplifier's input noise voltage en and offset voltage will be multiplied by,

$$\left(1 + \frac{Rf}{Rsh}\right).$$

This is superimposed on the output voltage Vout, and the operational amplifier's bias current error will also increase. It is therefore not practical to use an infinitely large Rf.

If there is an input capacitance Ct, the feedback capacitance Ct prevents high-frequency oscillations and also forms a lowpass filter with a time constant Cf × Rf value.

The value of Cf should be chosen according to the application. If the input light is similar to a discharge spark, and it is desired to integrate the amount of light, Rf can be removed so that the operational amplifier and Cf act as an integrating circuit.

However, a switch is required to discharge Cf before the next integration.

Bias Current

Since the actual input impedance of an operational amplifier is not infinite, some bias current that will flows into or out of the input terminals.

This may result in error, depending upon the magnitude of the detected current. The bias current, which flows in an FET, input operational amplifier is sometimes lower than 0.1 pA.

Bipolar operational amplifiers, however, have bias currents ranging from several hundred pA to several hundred nA.

However, the bias current of an FET operational amplifier increases two-fold for every increase of 5 to 10 °C in temperature, whereas that of bipolar amplifiers decreases with increasing temperature.

The use of bipolar amplifiers should be considered when designing circuits for high temperature operation. As is the case with offset voltage, the error voltage attributable to the bias current can be adjusted by means of a potentiometer connected to the offset adjustment terminals.

Furthermore, leakage currents on the PC board used to house the circuit may be greater than the operational amplifier's bias current.

Consideration must be given to the circuit pattern design and parts layout, as well as the use of Teflon terminals and guard rings.

Gain Peaking

The frequency response of a photodiode and operational amplifier circuit is determined by the time constant Rf × Cf. However, for large values of terminal capacitance (i.e. input capacitance) a phenomenon known as gain peaking will occur.

Figure below shows an example of such a frequency response. It can be seen from the figure that the output voltage increases sharply in the high frequency region, causing significant ringing

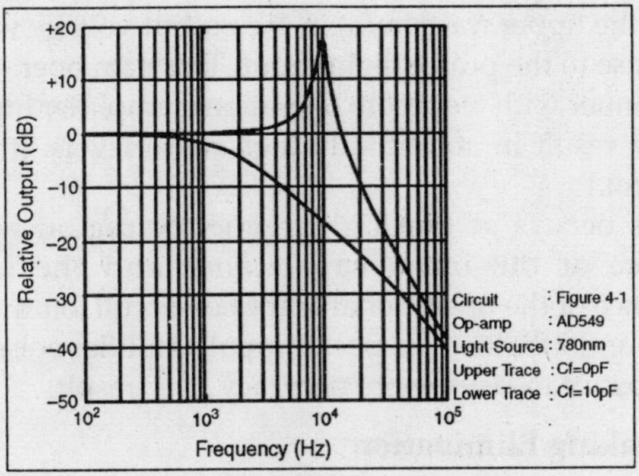

Fig. (a) Frequency response

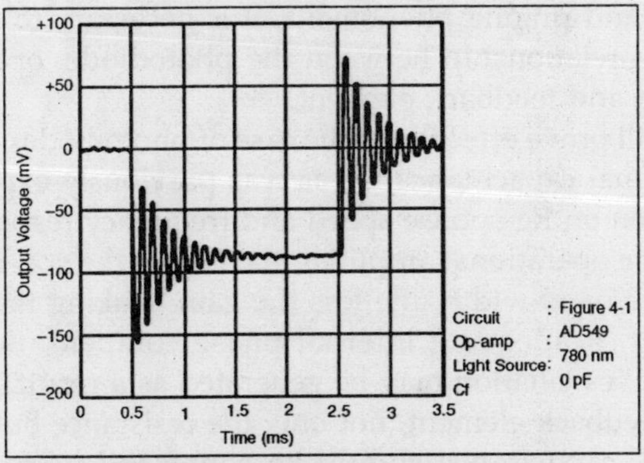

Fig. (b) Light pulse response

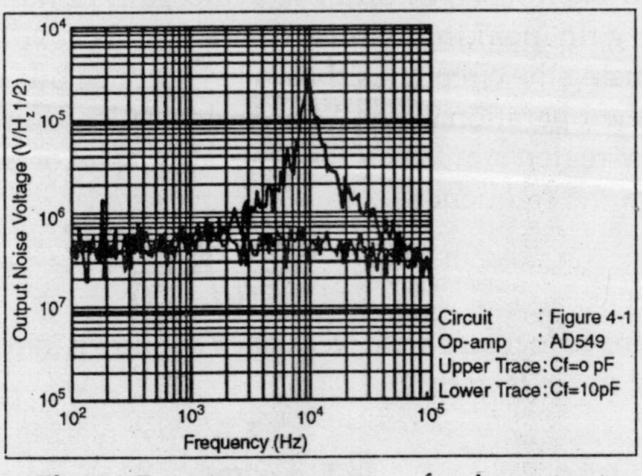

Fig. (c) Frequency response of noise output

See the upper trace in (a).in the output voltage waveform in response to the pulsed light input. This gain operates in the same manner with respect to operational amplifier input noise and may result in abnormally high noise levels. [the upper trace in (c).]

This occurs at the high frequency region when the reactance of the input capacitance and the feedback capacitance of the operational amplifier circuit jointly form an unstable amplifier with respect to input amplifier noise. In such a case, loss of measurement accuracy may result.

Gain Peaking Elimination

To achieve a wide frequency characteristic without gain peaking and ringing phenomena, it is necessary to select the optimum relationship between the photodiode, opera-tional amplifier and feedback element.

It will prove effective in the case of photodiodes to reduce the terminal capaci-tance Ct, as was previously explained in the section on Re-sponse speed and frequency response.

In the operational amplifier, the higher the speed and the wider the band-width, the less the gain peaking that occurs. However, if adequate internal phase compensation is not provided, os-cillation may be generated as a result.

A feedback element, not only the resistance but also the feedback capacitance should be connected in parallel, as explained previously, in order to avoid gain peaking.

The gain peaking phe-nomena can be explained as follows, using the circuit. As shown in Figure below, the circuit gain of the operational amplifier is determined for the low-frequency region simply by the resistance ratio of Rsh to Rf.

From the frequency,

$$f1 = \frac{Rsh + Rf}{2\pi RshRf(cf + Ct)}$$

gain begins to increase with frequency as shown in re-gion (2).

Next, at the frequency:

$$f2 = \frac{1}{2\pi CfRf}$$

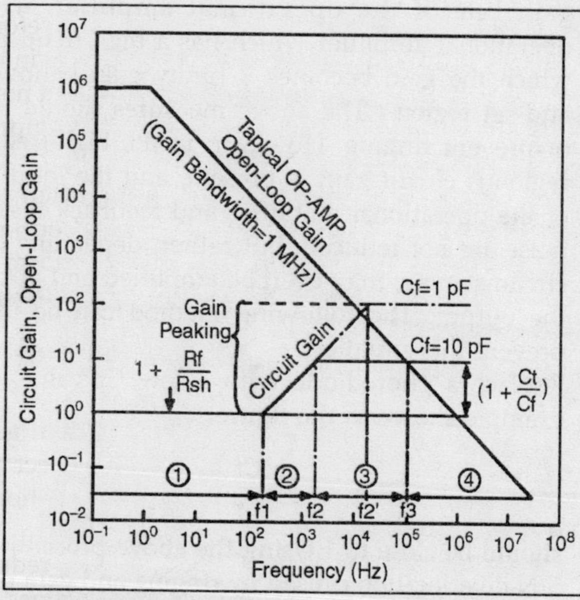

Fig. Graphical Representation of Gain Peaking

and above, the circuit gain of the operational amplifier enters a flat region (region (3)) which is determined by the ratio of Ct and Cf. At the point where frequency f3 intersects the open-loop gain frequency response at rolloff (6 dB/octave) of the operational amplifier, region is entered.

In this example, f1 and f2 correspond to 160 Hz and 1.6 kHz respectively under the conditions of Figure 4-1. If Cf is made 1 pF, f2 shifts to f2′ and circuit gain increases further.

What should be noted here is that, since the setting of increasing circuit gain in region *f* exceeds the open-loop gain curve, region (3) actually does not exist. As a result, ringing occurs in the pulsed light response of the operational amplifier circuit, and the gain peaking occurs in the frequency, then instability results.

To summarize the above points:

a) When designing Rf and Cf, f2 should be set to a value such that region *f* in Figure 4-3 exists.

b) When f2 is positioned to the right of the open-loop

gain line of the operational amplifier, use the operational amplifier which has a high frequency at which the gain becomes 1 (unity gain bandwidth), and set region *f*.The above measures should reduce or prevent ringing. However, in the high-frequency region *f*, circuit gain is present, and the input noise of the operational amplifier and feedback resistance noise are not reduced, but rather, depending on the circumstances, may even be amplified and appear in the output. The following method can be used to prevent this situation.

c) Replace a photodiode with a low Ct value. In the example shown in the figure,

$$\left(1+\frac{Ct}{Cf}\right)$$

should be close to 1.Using the above procedures, the S/N deterioration caused by ringing and gain peaking can usually be solved. However, regardless of the above measures, if load capacitance from several hundred pF to several nF or more, for example, a coaxial cable of several meters or more and a capacitor is connected to the operational amplifier output, oscillation may occur in some types of operational amplifiers. Thus the capacitance load must be set as small as possible.

SI APD

Advantage of APD

When using a opto-semiconductor for low-light-level measurement, it is necessary to take overall performance into account, including not only the opto-semiconductor characteristics but also the readout circuit (operational amplifier, etc.) noise. When a Si photodiode is used as a photodetector, the lowest detection limit is usually determined by the readout circuit noise because photodiode noise level is very low.

This tendency becomes more obvious when the higher frequency of signal to be detected. This is because the high-speed readout circuit usually exhibits larger noise, resulting in a predominant source of noise in the entire circuit system.

In such cases, if the detector itself has an internal gain mechanism and if the output signal from the detector is thus adequately amplified, the readout circuit can be operated so that its noise contribution is minimized to levels equal to one divided by gain (1/10 th to 1/100 th).In this way, when the lowest detection limit is determined by the readout circuit, use of an APD offers the advantage that the lowest detection limit can be improved by the APD gain factor to a level 1/10 th to 1/ 100 th of the lowest detection limit obtained with normal photodiodes.

Noise Characteristic of APD

When the signal is amplified, the inherent excess noise resulting from statistical current fluctuation current fluctuation in the avalanche multiplication process is also generated.

This noise current can be expressed by the following equation:

$$in = \sqrt{2qILM^2FB}.....(5-1)$$

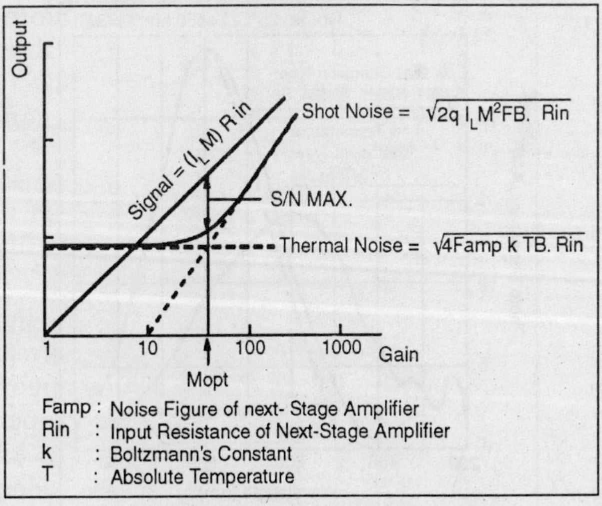

Fig. Noise Characteristic of APD

In the range of M=10 to 100, F is approximated M^x.
(*F*: Excess noise factor, M:Gain, IL: Photocurrentat at M=1),
(*q*: Electron charge, B: Bandwidth, x: Excess noise index)

In PIN photodiodes, using a large load resistance is not practical since it limits the response speed, so the circuit noise is usually dominated by the thermal noise of the photodiode.

In contrast, the gain of an APD, which is internally amplified, can be increased until the shot noise reaches the same level as the thermal noise. The APD can therefore offer an improved S/N without impairing the response speed.

Spectral Response of APD

The spectral response characteristics of the APD are almost the same as those of normal photodiodes if a bias voltage is not applied. When a bias voltage is applied, the spectral response curve will change.

This means that the gain changes depending on the incident light wavelength. This is because the penetration depth of light into the silicon substrate depends on the wavelength so that the wavelength absorption efficiency in the light absorption region differs depending on the APD structure. It is therefore important to select a suitable APD.

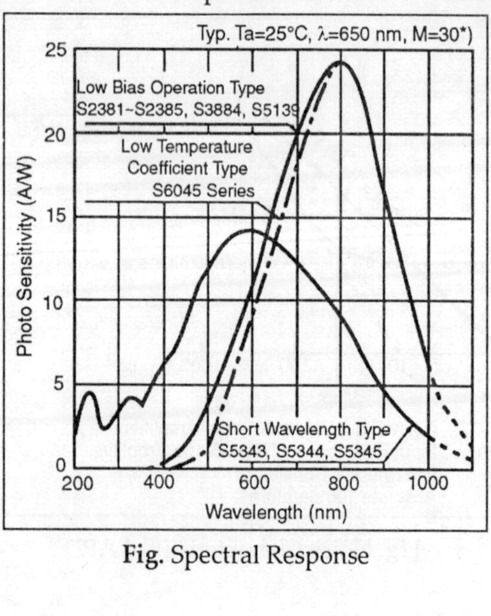

Fig. Spectral Response

To allow selection of spectral response characteristics, Hamamatsu provides two types of Si APDs: S2381 series and S6045 series for near infrared detection and S5343 series for light detection at shorter wavelengths. Figure below shows typical spectral response characteristics measured with a gain of 30 at 650 nm wavelength.

Temperature Characteristic of Gain

APD gain varies with temperature. For example, when an APD is operated at a constant bias voltage, the gain decreases with increasing temperature.

Therefore, in order to obtain a constant output, it is necessary to vary the bias voltage according to the APD temperature or to keep the APD at a constant temperature.In S2381 series, the temperature coefficient of the bias voltage is nearly equal to that of the breakdown voltage which is 0.65 V/°C Typ. at a gain of 100.Hamamatsu also provides S6045 series APDs which are designed to have an improved temperature coefficient (0.4 V/°C Typ.).

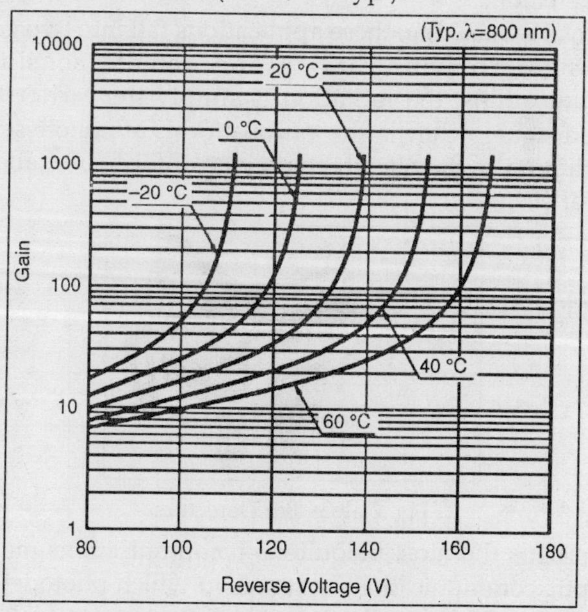

Fig. Gain Temperature Characteristics

PHOTO DETECTORS

INTRODUCTION

Photodetectors among the most ubiquitous types of technology in use today. They range from simple devices that automatically open supermarket doors, to receivers on TV and VCR remote controls, to photodiodes in a fiberoptic connection, to the CCD in a video camera, to enormous arrays used by astronomers to detect radiation from the other side of the universe.

Photodetectors are present in a huge variety of devices used in commerce, industry, entertainment and research. In fact, the field of photodetector design and use has grown to the point that few practitioners have a complete overview.

For our purposes, photodetectors include any device for registering photons with frequencies above that of radio waves—from far infrared on up to gamma rays. In this article we survey the main types of applications that use photodetectors.

Broadly speaking, these applications fall into two general groupings—communications and remote sensing. In communications, the radiation is simply the carrier for an encoded signal, while in the various forms of remote sensing, the radiation is the signal, conveying information about an object or scene.

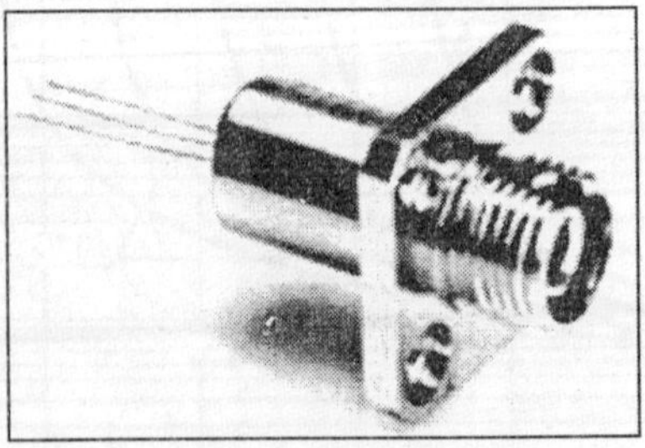

Fig. Galnas Pin Detectors

Perhaps the most ubiquitous communications market is fiberoptic communications systems, in which photodetectors, generally operating in the infrared (IR), pick up high-speed signals. These detectors do not need high sensitivity, since the laser drivers provide plenty of radiation to the fibre, but they must have extremely fast response, exhibit high reliability, and have a low cost. Photodiodes, especially those based on indium gallium arsenide (InGaAs) are the workhorses of optical communications, currently achieving data communication rates as high as 2.5 Gbits/s, which is more than 200,000 times the capacity of a single copper telephone wire.

While the detectors used in fiberoptic communications are invisible to the telephone user, those used in common remote control devices are obvious and are found in nearly every American home. Again, photodiodes sensitive to IR wavelengths are the standard, but requirements are far laxer than for fiberoptic devices, since data transmission rates are low, and only a small amount of information is generally conveyed—simple commands to change channels, or switch a VCR or TV on or off. Even remote controls, however, are increasing in sophistication—cordless mouse devices for PCs, for example, use detectors that measure the relative strength of signals to determine mouse orientation and position.

Safety and Security

The simplest types of remote sensing applications involve just the detection of the presence or absence of an object or a condition for safety and security monitoring. Such applications typically use photoconductors, the cheapest and most rugged of detector technologies. The most common of such applications is IR-sensitive motion detectors for home security systems. More recently developed examples include such automotive uses as collision detectors, to monitor objects in a driver's "blind spot," and passenger detectors that determine when to activate air bags. In factories, safety detectors sensitive to visible or ultraviolet (UV) wavelengths perform such tasks as electrical arc detection, automatically cutting off the current where arcing occurs.

Process Control

The next step up in sophistication comes in process control

applications, another large volume consumer application of photodetectors.

In many cases, these devices may be as simple as position sensors to check that a work piece is in the proper place or to provide feedback for robotic systems.

But other systems need accurate comparison of radiation intensity at different wavelengths, generally requiring accurately repeatable detectors such as photodiodes or phototransistors.

In recycling plants, for example, the natural fluorescence of certain plastics can be used with optically filtered detectors to sort different materials, while other spectroscopic tests can differentiate different types of glass.

Fig. 2 Germanium Position Sensors

Comparison of radiation intensity at two or more IR wavelengths is used in pyrometry to remotely measure and control high-temperature processes. Similar techniques in the UV are used to monitor and control flames. In some of these applications, in which radiation is intense, ruggedized bolometers—which measure radiation by the heat transferred, rather than by quantum detection of individual photons—are frequently preferred.

At the end of the production process, detectors are used as part of nondestructive testing units, using thermal scans to check for physical defects or for surface composition and finish. Increasingly, imaging systems, especially CCDs, are used in machine vision systems that are incorporated into more flexible robots, not only in factories but also, to a limited extent, in mobile devices used in service institutions, such as hospitals.

Process control applications in factories often require detector systems that can endure extreme environmental conditions.

In the food industry, for example, sanitizing caustic solutions are sprayed out at high temperature and at pressures up to 1200 psi, necessitating extremely durable photodetector packaging. In metalworking plants, detectors may require Teflon housing to facilitate removal of molten metal splatter from spot welding. Such environments also put a premium on detector materials able to withstand wide ranges in temperature conditions to reduce insulation needs.

A third high-volume detector application is in all forms of video cameras, from the large cameras used by broadcast and cable TV studios to a palm-sized camcorder. While this is a relatively mature field, refinements in light sensitivity, response time, and resolution are continuing in the basic CCD technology that dominates the video camera business (see Fig). The latest development in the field involves the extension of mass-production video camera technology to IR night-vision applications that were previously limited mostly to defence uses.

THE CUTTING EDGE

The most advanced detectors tend to be made for low-volume specialized fields, such as laboratory research, biomedicine, defence, environmental monitoring, and astronomy.

Thermography, measuring the heat radiated by the human body, has long been a routine diagnostic test, requiring IR detectors. Recent developments have allowed research biologists to observe rapid biochemical reactions with ultrahigh-speed CCDs.

Cameras used for tracing chemical diffusion in individual cells, for example, typically have read-out rates of several megahertz and frame rates of up to 10,000 per second. In spectroscopic applications, the high sensitivity of photomultiplier tubes (PMTs) makes them the workhorse of laboratory detectors.

Undoubtedly the biggest drivers of high-technology detectors, especially in the IR, have been defence applications. For decades, the armed forces have used IR detectors on heat seeking missiles, thus setting off a long-term arms race, as decoys such as magnesium flares confused these detectors, prompting the development of more sophisticated devices to distinguish the decoys from real targets.

This has led to the development of tunable IR detectors that can pick out jet exhaust spectra, satellite-borne staring array detectors to isolate warm targets against a warm background, and continuous improvements in IR video cameras for night vision.

Infrared scanning from aircraft reveals the location of land mines, which show up as regions of warmer soil. The ultimately futile missile defence programme, the Strategic Defence Initiative (SDI), while it never arrived at a practical way to shoot down ICBMs, did lead to the development of sensitive far-IR detectors for wavelengths longer than 5 μm, which subsequently found use in astronomy, among other places.

With the scaling back of SDI and the end of the Cold War, many military technologies are moving out into civilian applications. Work is well advanced on adapting military night vision IR detector systems to enhance driver vision beyond the reach of headlights. Workhorses of low-light detection such as avalanche photodiodes first emerged from Pentagon-funded research.

USES

Environmental Sensing

Environmental monitoring today uses a broad range of photodetectors from the UV to the IR. Typically, signals are low intensity, so the primary detectors are PMTs and avalanche photodiodes (solid-state photomultipliers). Pollution detection generally relies on UV spectroscopy, with detectors measuring the strength of absorption lines for such pollutants as sulfur dioxide, nitrous oxides, and ozone.

Fluorescence spectroscopy allows detection of extremely small amounts of pollutants such as benzene, toluene, xylene and sulfur oxides, which can be measured at a parts-per-billion level. Another sensitive technique is chemiluminescence, in which a reactant is introduced that selectively combines with a pollutant, producing a faint luminescence in the process. Large active areas and low noise levels are required for detectors in such applications.

More straightforward techniques are used for monitoring solid particulates in air and water. Here, the amount of light scattered by the particles is a good continuous measure of pollution levels. Ample light is often available, so photodiodes can be used, taking advantage of their robustness and fast response times.

For monitoring of pollution over large areas, lidar scattering techniques are increasingly used. In this method, a laser sends out light pulses at a repetition rate of about 50 Hz; these pulses scatter off pollutants in the air.

Sensitive detectors, generally PMTs, record the amount of light scattered and the time of arrival of the scattered light. In this way, the distance to the pollutant can be measured. As the lidar scans across a region, the distribution of a given pollutant can be mapped.

Space-based environmental monitoring, both for tracking pollutants and for weather and climate analysis make greater demands on detectors, both for sensitivity and for compactness.

Satellite-mounted IR spectrometers can measure ozone concentrations from space, sometimes using pyroelectric detectors to convert heat into electric impulses. Space-based IR detectors for temperature monitoring are generally mercury-cadmium-telluride IR arrays that were first developed for defence purposes.

Astronomy

Without doubt the most challenging applications for photodetectors come in the field of astronomy, where the range of wavelengths studied extends from the far-IR at hundreds of microns to cosmic ray photons with 1020 eV of energy and wavelengths of 10-20 μm, a range of 22 orders of magnitude.

At the long wavelength end, the European Space agency is making plans for FIRST, the far-infrared and submillimeter space telescope, which will open up the one spectral band not yet surveyed—from 60 to 670 μm. The design will use arrays of germanium gallium photo-conductors for 100 to 200 μm and arrays of bolometers cooled to 0.1 K for the rest of the band.

At near-IR wavelengths, focal plane arrays of indium antinomide, mercury cadmium telluride, and platinum silicide, all initially developed for the military, are becoming routing accessories for both ground- and space-based telescopes.

Miniaturized IR and CCD cameras on board the spacecraft Clementine enabled it to survey the moon at wavelengths ranging from the UV to 9.5 μm in the far-IR.

Fig Ultraminiature, High-resolution CCD Colour board Video camera

At UV wavelengths, necessarily viewed from space because the atmosphere is mostly opaque to UV, advances in microchannel-plate PMTs have led to major improvements in sensitivity, evident in such instruments as those used in the ASTRO platform on board the space shuttles.

At still higher energies, x-ray telescopes such as ROSAT, launched by the European Space Agency, are opening new windows into the most energetic processes in the universe. X-ray detectors generally are based on approaches different from those used for longer wavelength radiation.

In gas-filled detectors, electrons are knocked out by x-rays, and electron multiplication occurs in the gas as electrons avalanche toward a high-voltage wire. However, techniques borrowed from other parts of the spectrum, such as microchannel plates for electron multiplication, can be used by x-ray detectors as well.

For soft x-rays, below 10 keV, specially designed CCDs are finding greater application. Higher energy detectors often employ calorimeters, which measure the total heat liberated when x-rays are absorbed. And research is currently developing fundamentally new types of detectors based on superconducting tunnel junctions.

TYPES

Photovoltaic Effect Detector

The *photovoltaic effect* occurs at a junction in a semiconductor. The junction is the boundary between a region where the conductivity is due to electrons and a region where the conductivity is due to holes (the absence of electrons). This is called a pn junction.

At the junction, an electric field is present internally because there is a change in the level of the conduction and valence bands. This change leads to the familiar electrical rectification effect produced by such junctions.

The photovoltaic effect is the generation of a voltage when light strikes a semiconductor pn junction. The photovoltaic effect is measured using a high-impedance voltage-measuring device, which essentially measures the open-circuit voltage produced at the junction.

In the dark, no open circuit voltage is present. When light falls on the junction, the light is absorbed and, if the photon energy is large enough, it produces free hole-electron pairs. The electric field at the junction separates the pair and moves the electron into the n-type region and the hole into the p-type region.

This leads to an open circuit voltage that can be measured externally. This process is the origin of the photovoltaic effect. We note that the open-circuit voltage generated in the photovoltaic effect may be detected directly and that no bias voltage nor load resistor is required.

If the junction is short-circuited by an external conductor, current will flow in the circuit when the junction is illuminated. One may measure either the open-circuit voltage or the short-circuit current. Both these quantities will give measures of the light falling on the junction.

Photoemissive Effect Detector

Now we turn to the *photoemissive effect*. The photoemissive effect involves the emission of electrons from a surface irradiated by quanta of light energy.

A photoemissive detector has a cathode coated with a material that emits electrons when light of wavelength shorter than the cutoff wavelength falls on the surface.

The electrons emitted from the surface are accelerated by a voltage to an anode, where they produce a current in an external circuit.

The detectors are enclosed in a vacuum environment to allow a free flow of electrons from anode to cathode. These detectors are available commercially from a number of manufacturers. They represent an important class of detectors for many applications.

Some spectral response curves for photoemissive cathodes are shown in Figure below. The cathodes are often mixtures containing alkali metals, such as sodium and potassium, from which electrons can easily be emitted.

The responsivity in mA/watt of these devices is shown in the figure from the ultraviolet to the near infrared. At wavelengths longer than about 1000 nm, no photoemissive response is available.

The short-wavelength end of the response curve is set by the nature of the window material used in the tube that contains the detector. The user can select a device that has a cathode with maximum response in a selected wavelength region. An important variation of the photoemissive detector is the photomultiplier, which will be described later.

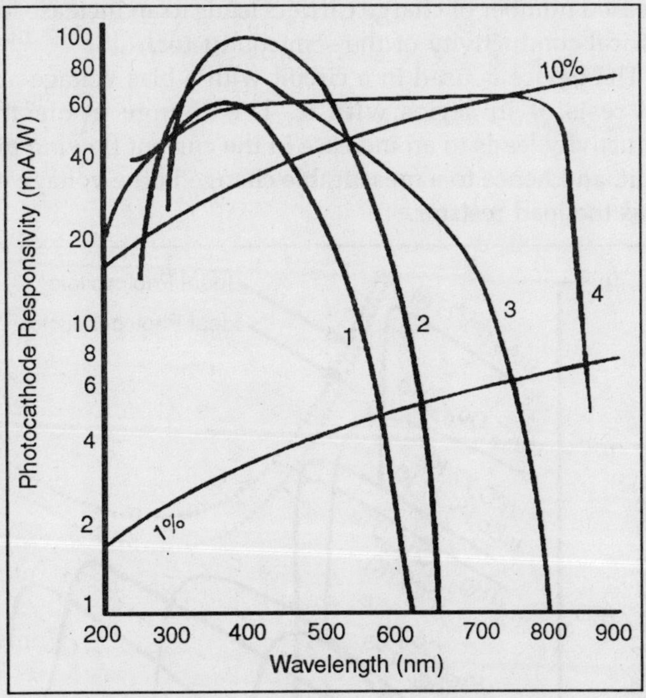

Fig. Response as a Function of Wavelength for a Number of Photoemissive Surfaces.

Curve 1 is the response of a bialkali type of cathode with a sapphire window; curve 2 is for a different bialkali cathode with a lime glass window; curve 3 is for a multialkali cathode with a lime glass window; and curve 4 is for a GaAs cathode with a 9741 glass window. The curves labeled 1% and 10% denote what the response would be at the indicated value of quantum efficiency.

Photoconductivity Detector

A third phenomenon used in optical detectors is *photoconductivity*. A semiconductor in thermal equilibrium contains free electrons and holes. The concentration of electrons and holes is changed if light is absorbed by the semiconductor. The light must have photon energy large enough to produce free electrons within the material. The increased number of charge carriers leads to an increase in the electrical conductivity of the semiconductor.

The device is used in a circuit with a bias voltage and a load resistor in series with it. The change in electrical conductivity leads to an increase in the current flowing in the circuit, and hence to a measurable change in the voltage drop across the load resistor.

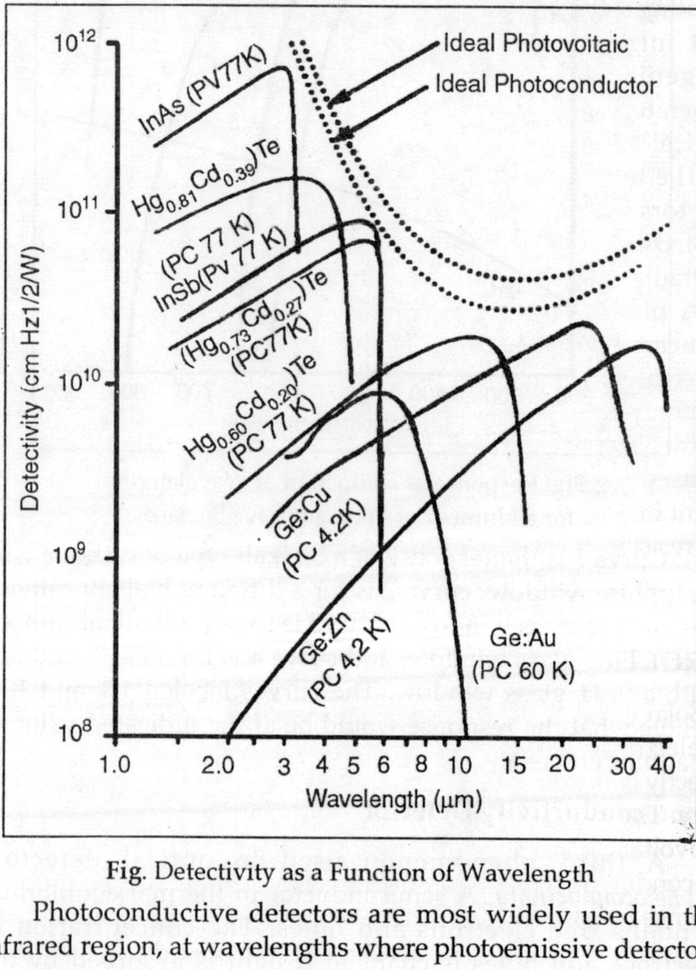

Fig. Detectivity as a Function of Wavelength

Photoconductive detectors are most widely used in the infrared region, at wavelengths where photoemissive detectors are not available. Many different materials are used as infrared photoconductive detectors.

Some typical values of detectivity (in cm-Hz$^{1/2}$/watt) as a function of wavelength for some devices operating in the infrared are shown in Figure below, along with values of detectivity for some other detectors to be discussed later. The photoconductive detectors are denoted PC. The exact value of detectivity for a specific photoconductor depends on the operating temperature and on the field of view of the detector. Most infrared photoconductive detectors operate at a cryogenic temperature (frequently liquid nitrogen temperature, 77 K) which may involve some inconvenience in practical applications.

The temperature of operation is indicated. Photovoltaic detectors are denoted PV; photoconductive detectors are denoted PC. The curves for ideal photodetectors assume a 2psteradian field of view and a 295 K background temperature.

A photoconductive detector uses a crystal of semiconductor material that has low conductance in the dark and an increased value of conductance when it is illuminated. It is commonly used in a series circuit with a battery and a load resistor. The semiconductor element has its conductance increased by light. The presence of light leads to increased current in the circuit and to increased voltage drop across the load resistor.

SOLAR CELL

INTRODUCTION

Solar cells are devices which convert solar energy directly into electricity, either directly via the photovoltaic effect, or indirectly by first convertin the solar energy to heat or chemical energy. The most common form of solar cells are based on the photovoltaic (PV) effect in which light falling on a two layer semi-conductor device produces a photovoltage or potential difference between the layers. This voltage is capable of driving a current through an external circuit and thereby producing useful work.

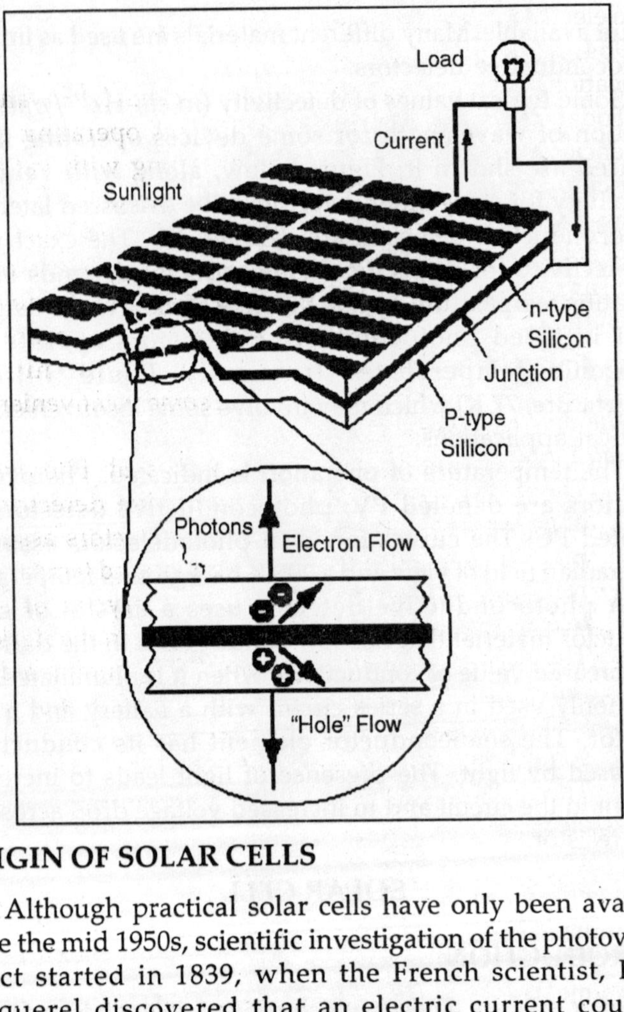

ORIGIN OF SOLAR CELLS

Although practical solar cells have only been available since the mid 1950s, scientific investigation of the photovoltaic effect started in 1839, when the French scientist, Henri Becquerel discovered that an electric current could be produced by shining a light onto certain chemical solutions.

The effect was first observed in a solid material (in this case the metal selenium) in 1877. This material was used for many years for light meters, which only required very small amounts of power. A deeper understanding of the scientific principles, provided by Einstein in 1905 and Schottky in 1930, was required before efficient solar cells could be made. A silicon solar cell which converted 6% of sunlight falling onto it into electricity was developed by Chapin, Pearson and Fuller in 1954, and this kind of cell was used in specialised applications such as orbiting space satellites.

Today's commercially available silicon solar cells have efficiencies of about 18% of the sunlight falling on to them into electricity, at a fraction of the price of thirty years ago. There is now a variety of methods for the practical production of silicon solar cells (amorphous, single crystal, polycrystalline), as well as solar cells made from other materials (copper indium diselenide, cadmium telluride, etc).

NEED FOR SOLAR CELL

The development of solar cell use in Australia has been stimulated by:

- The need for low maintenance, long lasting sources of electricity suitable for places remote from both the main electricity grid and from people; eg satellites, remote site water pumping, outback telecommunications stations and lighthouses;
- The need for cost effective power supplies for people remote from the main electricity grid; eg Aboriginal settlements, outback sheep and cattle stations, and some home sites in grid connected areas.
- The need for non polluting and silent sources of electricity; eg tourist sites, caravans and campers
- The need for a convenient and flexible source of small amounts of power; eg calculators, watches, light meters and cameras;
- The need for renewable and sustainable power, as a means of reducing global warming.

Together, these needs have produced a growing market for photovoltaics which has stimulated innovation. As the market has grown, the cost of cells and systems has declined, and new applications have been discovered.

HOW ARE SOLAR CELLS MADE?

Silicon solar cells are made using either single crystal wafers, polycrystalline wafers or thin films.

Single crystal wafers are sliced, (approx. 1/3 to 1/2 of a millimetre thick), from a large single crystal ingot which has been grown at around 1400 ýÿC, which is a very expensive process. The silicon must be of a very high purity and have a near perfect crystal structure (see figure (a)).

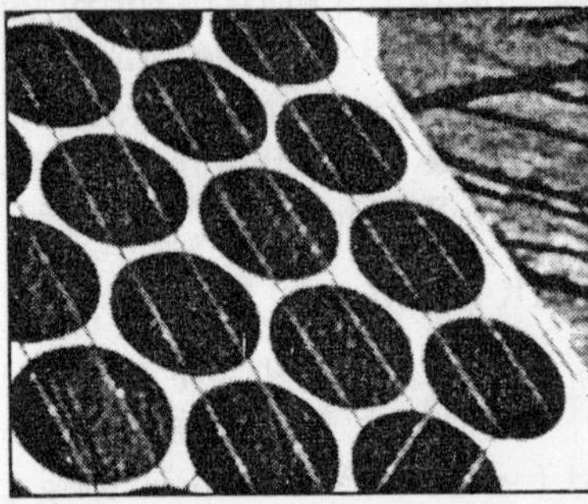

Fig. (a) Single Crystal Solar Cells in panel

Fig. (b) Polycrystalline solar panel

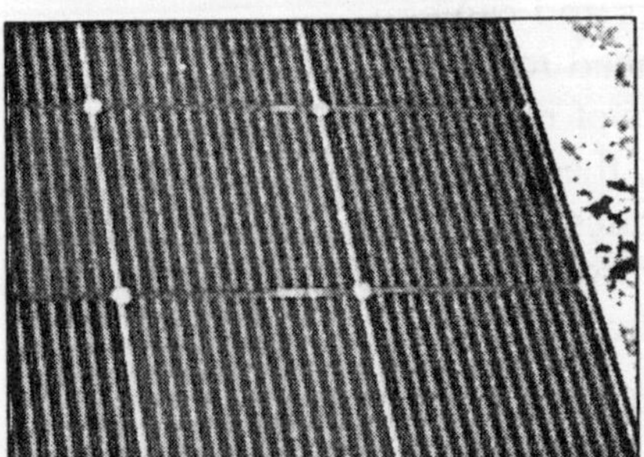

Fig. (c) a-Si solar panel

Figure Different types of Silion solar cells:

Polycrystalline wafers are made by a casting process in which molten silicon is poured into a mould and allowed to set. Then it is sliced into wafers (see figure (b)).As polycrystalline wafers are made by casting they are significantly cheaper to produce, but not as efficient as monocrystalline cells. The lower efficiency is due to imperfections in the crystal structure resulting from the casting process.

Almost half the silicon is lost as saw dust in the two processes mentioned above. Amorphous silicon, one of the thin film technologies, is made by depositing silicon onto a glass substrate from a reactive gas such as silane (SiH_4) (see figure (c)). Amorphous silicon is one of a number of thin film technologies.

This type of solar cell can be applied as a film to low cost substrates such as glass or plastic. Other thin film technologies include thin multicrystalline silicon, copper indium diselenide/ cadmium sulphide cells, cadmium telluride/cadmium sulphide cells and gallium arsenide cells.

There are many advantages of thin film cells including easier deposition and assembly, the ability to be deposited on inexpensive substrates or building materials, the ease of mass production, and the high suitability to large applications.

In solar cell production the silicon has dopant atoms introduced to create a p-type and an n-type region and thereby producing a p-n junction. This doping can be done by high temperature diffusion, where the wafers are placed in a furnace with the dopant introduced as a vapour.

There are many other methods of doping silicon. In the manufacture of some thin film devices the introduction of dopants can occur during the deposition of the films or layers.

A silicon atom has 4 relatively weakly bound (valence) electrons, which bond to adjacent atoms. Replacing a silicon atom with an atom that has either 3 or 5 valence electrons will therefore produce either a space with no electron (a hole) or one spare electron that can move more freely than the others, this is the basis of doping. P-type doping, the creation of excess holes, is achieved by the incorporation into the silicon of atoms with 3 valence electrons, most often boron and n-type doping, the creation of extra electrons is achieved by incorporating an atom with 5 valence electrons, most often phosphorus (see figure).

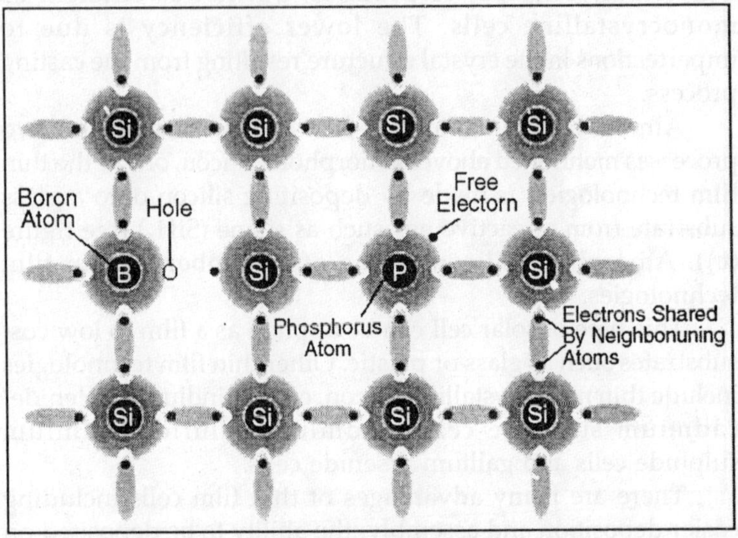

Fig. Silicon Crystal Lattice with Dopant Atoms.

WORKING

Solar cells are made using semiconductors such as silicon. Semiconductors have interesting electrical properties, making them useful for electronic devices such as the microprocessors used in computers. One of their properties is that they can be treated in different ways to become either 'positive' (p-type) or 'negative' (n-type).

A solar cell consists of two layers of semiconductor, one p-type and the other n-type, sandwiched together to form a 'pn junction'.

This pn interface induces an electric field across the junction. When particles of light ('photons') are absorbed by the semiconductor, they transfer their energy to some of the semiconductor's electrons, which are then able to move about through the material. For each such negatively charged electron, a corresponding mobile positive charge, called a 'hole', is created. In an ordinary semiconductor, these electrons and holes recombine after a short time and their energy is wasted as heat.

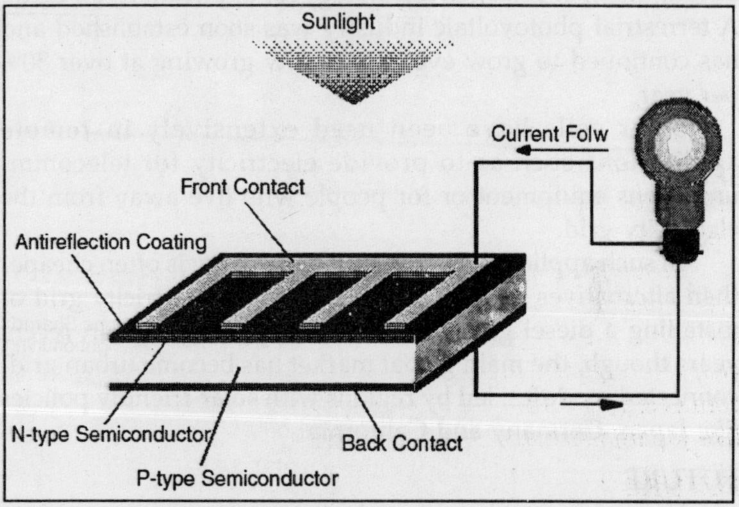

Fig. Photovoltaic Cell Under Illumination

In a solar cell, however, the electrons and holes near the pn junction are swept across in opposite directions by the action of the electric field and others diffuse towards the junction to replace them.

This separation of charge induces a voltage across the device. By connecting the device to an external circuit, the electrons are able to flow - and this flow of electrons is what we call electricity.

WHAT ARE SOLAR CELLS GOOD FOR?

In the mid-1950s, the builders of early space craft required an energy source that was reliable, long lasting and required no maintenance.

There are no power points in space and, as we all know, batteries can go flat pretty quickly. Solar cells, however, are an ideal choice for this application. They are reliable, maintenance-free and their energy source-sunlight is abundant and virtually everlasting. The first practical solar cells were therefore developed for these space applications.

Unfortunately, these first solar cells were very expensive to produce. In the early 1970s, the oil crisis spurred development of solar cell technology for applications on earth. A terrestrial photovoltaic industry was soon established and has continued to grow ever since, now growing at over 30% per year.

Solar cells have been used extensively in remote applications such as to provide electricity for telecommunications equipment or for people who live away from the electricity grid.

In such applications, solar cell technology is often cheaper than alternatives such as connecting to the electricity grid or installing a diesel generator and transporting fuel. In recent years though, the main global market has become urban grid-connected modules, led by regions with solar-friendly policies like Japan, Germany and California.

FUTURE

Recent developments suggest that photovoltaic technology may soon be playing a much larger role in all our lives. Spearheading this push has been the School of Photovoltaic and Renewable Energy Engineering.

While efficiencies have been improved, the cost of the solar cells themselves has steadily fallen with increasing mass production in recent years and should continue to do so over the coming decade.

Development of systems technology and integration of solar cells into building materials such as roof tiles is also reducing the overall system costs and making them more attractive to buyers.

As costs continue to fall over the coming decade, we are likely to see solar cells become more commonplace in everyday applications such as on people's roofs. Those who elect to place solar cells on their roof can sell electricity to their local utility during the day when the sun is shining and buy it back at night when it is not. This is the key to the boom in the market in Europe and spectacularly in Germany. These changes should have a significant impact on reducing the emission of greenhouse gases and other pollutants, and will take us one step closer to a cleaner energy future.

LIGHT EMITTING DIODES

INTRODUCTION

An LED is a special semiconductor which emits light when current is passed through it. There are many different physical styles. The emitted colour spectrum is usually very narrow.

Also Light-emitting diodes (LEDs) are solid-state devices that convert electric energy directly into light of a single colour. Because they employ "cold" light generation technology, in which most of the energy is delivered in the visible spectrum, LEDs don't waste energy in the form of non-light producing heat. In comparison, most of the energy in an incandescent lamp is in the infrared (or non-visible) portion of the spectrum. As a result, both fluorescent and HID lamps produce a great deal of heat. In addition to producing cold light,

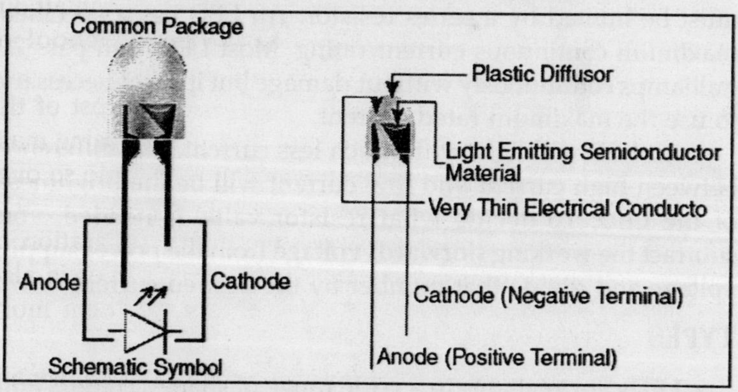

It can generally be specified as a specific wavelength in the electromagnetic spectrum. The emitted colour selection is somewhat limited. The most commonly available colors are red, green, amber, yellow, blue and white.

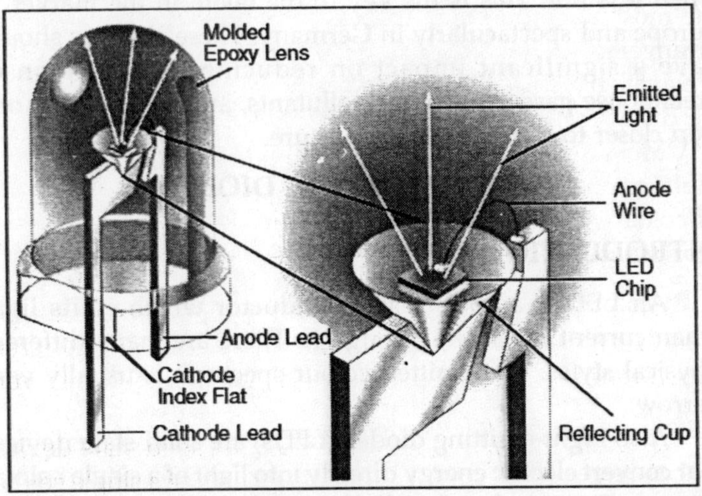

The red, green, yellow and amber have a working voltage of approximately 1.8 volts. The actual working voltage is determined by the breakdown voltage of the particular semiconductor material.

When using an LED in a circuit, the exact working voltage is not extremely important. The most important thing is the current flow through the LED. The current through the diode must be limited by a series resistor. An LED has a specified maximum continuous current rating. Most LEDs can pass 20 milliamps continuously without damage but it is not necessary to use the maximum rated current.

An LED will light with much less current. The difference between high current and low current will be the brightness of the LED. To decide what resistor value is needed, you subtract the working (forward) voltage from the power supply voltage and divide that number by the desired current flow.

TYPES

LEDs are available in a wide range of shapes, colours and various sizes with different light output intensities available, with the most common (and cheapest to produce) being the standard 5mm Red LED. LED's are also available in various "packages" arranged to produce both letters and numbers with the most common being that of a "Seven-Segment Display" arrangement. Nowadays, full colour flat screen LED displays are available with a large number of dedicated I.C.s available for driving the displays directly.

Most LEDs produce a single output colour however, multi-colour LEDs are now available that can produce a variety of colours within a single device. These are actually 2 LEDs fabricated within a single package as shown below.

A Bi-colour LED

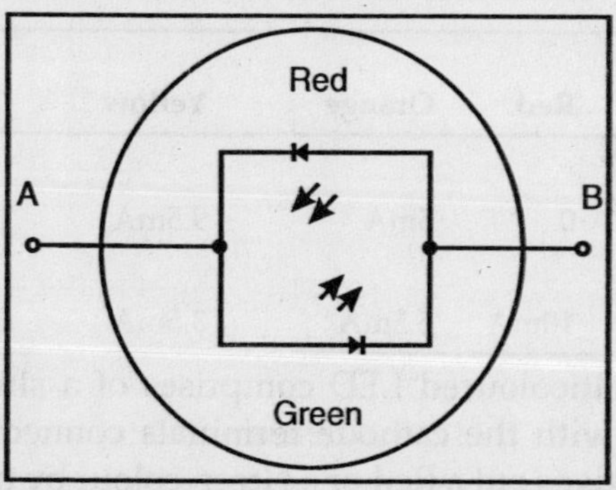

LED Selected	**Terminal A**		**AC**
	+	-	
LED 1	ON	OFF	ON
LED 2	OFF	ON	ON
Colour	Green	Red	Yellow

Here the LEDs are connected in "inverse parallel", so that the colour red is emitted when the device is connected in one direction and the colour green is emitted when it is biased in the other direction.

This type of arrangement is useful for giving polarity indication, for example correct connection of batteries, power supplies etc.

Also, both LEDs would take it in turn to light if the device was connected (via suitable resistor) to a low voltage, low frequency AC supply.

A Multi or Tri-colour LED

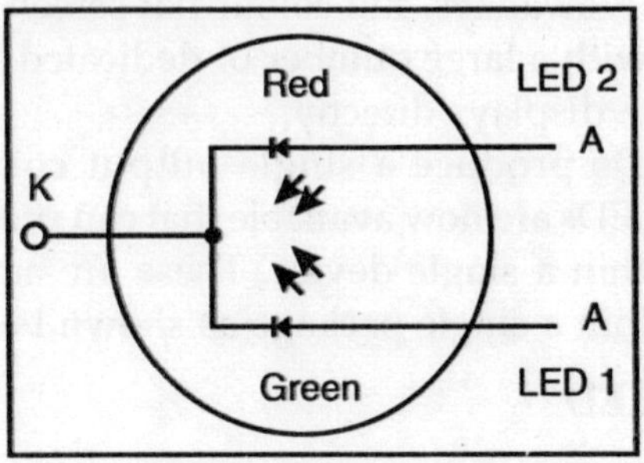

Output Colour	Red	Orange	Yellow	Green
LED 1 Current	0	5mA	9.5mA	15mA
LED 2 Current	10mA	6.5mA	3.5mA	0

This multicoloured LED comprises of a single Red and Green LED with the cathode terminals connected together. This device gives out a Red or a Green colour by turning "ON" only one LED at a time.

It can also generate additional shades of colours such as Orange or Yellow by turning "ON" the two LEDs in different ratios of forward current as shown in the table thereby generating 4 different colours from just two diode junctions.

PRINCIPLE

At initialization, the voltage applied to the junction is 2 volts (an arbitrary setting), and holes from the p-doped semiconductor region (GaAlAs) meet electrons from the n-doped region (GaAs) in the Active Layer to produce infrared light. In order to operate the tutorial, use the mouse cursor to translate the Voltage slider between a range of zero and four volts.

As the voltage is increased, the number of holes and electrons entering the junction region increases, as does the amount of light (number of colored photons) being emitted when the holes and electrons combine. The composition of the materials comprising the doped semiconductor diode can be altered with the LED Material slider.

As the slider is translated, the n-doped and p-doped materials change to new semiconductors that yield a variety of visible light wavelengths when electrons and holes meet and combine in the Active Layer region. The common name for each colour of light produced by the diode is presented above the LED Material slider bar.

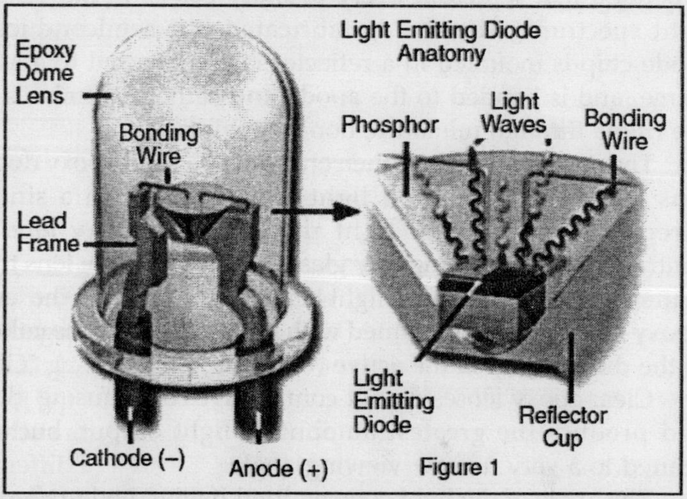

Figure 1

At the heart of a light emitting diode is a semiconductor chip fabricated in a manner similar to that of common integrated circuits, which are utilized a wide array of devices ranging from computers to household appliances.

The diode semiconductor contains several very thin layers of material that are sequentially deposited onto a supporting substrate (usually silicon or doped gallium). In the tutorial, the first semiconductor material that is deposited onto the substrate is doped with atoms containing excess electrons to yield what is termed an n-type semiconductor.

A second doped material, containing atoms having too few electrons (a p-type semiconductor), is then deposited onto the first semiconductor to form the diode. The region created between the doped semiconductor materials is known as the active layer. When a voltage is applied to the diode, holes (positive charges) from the p-type region and electrons (negative charges) from the n-type region meet in the active layer to produce light.

The wavelength of light emitted by the diode is dependent upon the chemical composition and relative energy levels of the doped semiconductor materials that comprise the p and n regions. By varying the composition of the doped semiconductors, a wide range of emitted wavelengths can be generated that represent every primary colour in the visible light spectrum. After being fabricated, the semiconductor diode chip is mounted in a reflector cup connected to a lead frame, and is bonded to the anode and cathode terminals of the frame through miniature bonding wires .

The entire assembly is then encased in a solid epoxy dome lens that enables emitted light to be focused in a single direction. Diffusion of light through the epoxy lens is controlled by embedding tiny glass particles into the lens that scatter light and spread the light beam emerging from the lens. Epoxy lenses can also be tinted with dyes to indicate the colour of the diode when in the active (on) state.

Clear epoxy lenses do not contain dyes or diffusing glass and produce the greatest amount of light output, but are limited to a very narrow viewing angle.

The angle of the light cone emitted by the diode-reflector cup combination can be altered by changing the epoxy lens shape, the geometry of the reflector cup, and the size and distance between the semiconductor diode and the nose of the epoxy lens. By varying the composition of the doped semiconductor materials utilized to fabricate the diode, common light emitting diodes can manufactured that emit light in both the visible and infrared wavelength regions.

The most common semiconductors employed in diode construction are aluminum, arsenic, gallium, indium, phosphorous and nitrogen (as nitrides), which can be combined in a variety of different proportions and doping levels. The wavelength (colour) of light produced by a light emitting diode is determined exclusively by the nature of the doped semiconductor materials and is independent of dyes utilized to colour the epoxy dome lens.

Typical visible wavelengths emitted by semiconductor diodes are red (650 nanometers), orange (620 nanometers), yellow (585 nanometers), green (555 nanometers), and blue (480 nanometers). White light LEDs are manufactured by a variety of mechanisms, but often contain a phosphor material in the reflector cup that intercepts high-energy blue light from the diode and emits secondary radiation across the entire visible light spectrum.

I-V CHARACTERISTICS

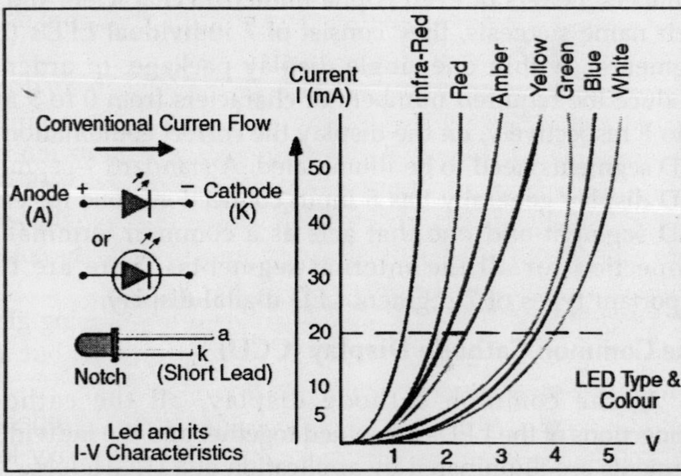

Before a light emitting diode can "emit" any form of light it needs a current to flow through it, as it is a current dependant device. As the LED is to be connected in a forward bias condition across a power supply it should be *Current Limited* using a series resistor to protect it from excessive current flow.

From the table above we can see that each LED has its own forward voltage drop across the PN-junction and this parameter which is determined by the semiconductor material used is the forward voltage drop for a given amount of forward conduction current, typically for a forward current of 20mA. In most cases LEDs are operated from a low voltage DC supply, with a series resistor to limit the forward current to a suitable value from say 5mA for a simple LED indicator to 30mA or more where a high brightness light output is needed.

LED DISPLAYS

As well as individual colour or multi-colour LEDs, light emitting diodes can be combined together in a single package to produce displays such as bargraphs, strips, arrays and 7-segment displays.

A 7-segment LED display provides a very convenient way of displaying information or digital data in the form of Numbers, Letters or even Alpha-numerical characters and as their name suggests, they consist of 7 individual LEDs (the segments), within one single display package. In order to produce the required numbers or characters from 0 to 9 and A to F respectively, on the display the correct combination of LED segments need to be illuminated. A standard 7-segment LED display generally has 8 input connections, one for each LED segment and one that acts as a common terminal or connection for all the internal segments. There are two important types of 7-segment LED digital display.

The Common Cathode Display (CCD)

In the common cathode display, all the cathode connections of the LEDs are joined together and the individual segments are illuminated by application of a HIGH, logic "1" signal.

The Common Anode Display (CAD)

In the common anode display, all the anode connections of the LEDs are joined together and the individual segments are illuminated by connecting the terminals to a LOW, logic "0" signal.

Typical 7-segment Display

A semiconductor-based laser used to generate analog signals or digital pulses for transmission through optical fibers. Both laser diodes and LEDs (light-emitting diodes) are used for this purpose, but the laser diode generates a smaller beam that is easier to couple with the smaller core of singlemode fibers.

Laser diodes are designed to emit light either from their edge or their surface, the latter providing a circular beam that couples better with the round core of the fibre.

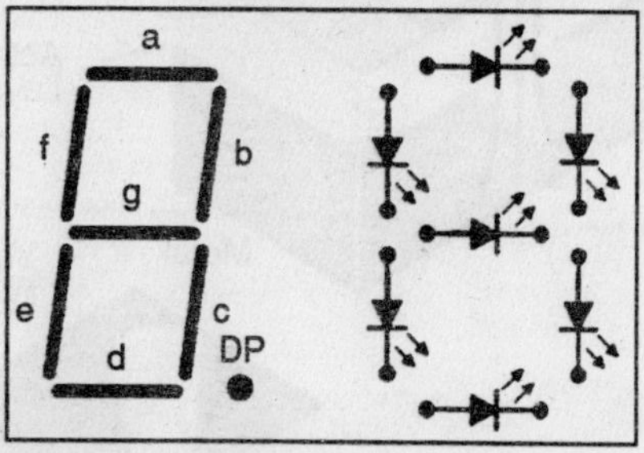

SEMICONDUCTOR LASERS

Laser diodes work on the same principle as the bigger gas lasers. They function as an optical oscillator by stimulating a chain reaction of photon emission inside a tiny chamber.

In edge-emitting lasers, the semiconductor waver is cleaved, and the inherent properties of the semiconductor create reflective ends that may or may not be enhanced with additional reflective films. With vertical cavity surface emitting lasers (VCSELs), the reflectivity has to be added. The most common semiconductors used in laser diodes are compounds based on gallium arsenide (750 to 900 nm in the infrared), indium gallium arsenide phosphide (1200 to 1700 nm in the infrared) and gallium nitride (near 400 nm in the blue).

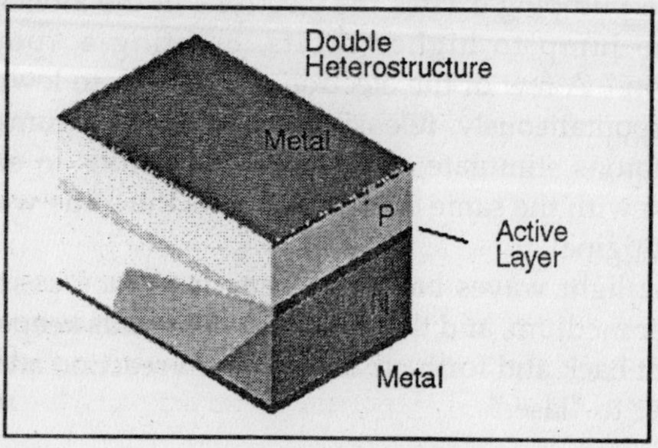

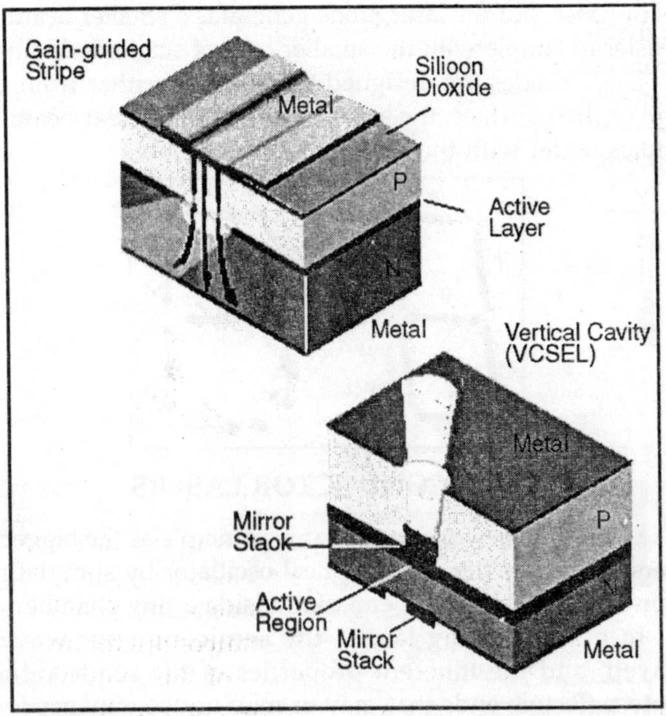

WORKING

A laser is an optical oscillator, which is made out of a solid, liquid or gas with mirrors at both ends. To make the laser work, the material is excited or "pumped," with light or electricity.

The pumping excites the electrons in the atoms, causing them to jump to higher orbits, creating a "population inversion." A few of the electrons drop back to lower energy levels spontaneously, releasing a photon (quantum of light). The photons stimulate other excited electrons to emit more photons with the same energy and thus the same wavelength as the original.

The light waves build in strength as they pass through the laser medium, and the mirrors at both ends keep reflecting the light back and forth creating a chain reaction and causing the laser to "lase."

In simple laser cavities, one mirror has a small transparent area that lets the laser beam out. In semiconductor lasers, both mirrors often transmit a beam, the second one being used for monitoring purposes.

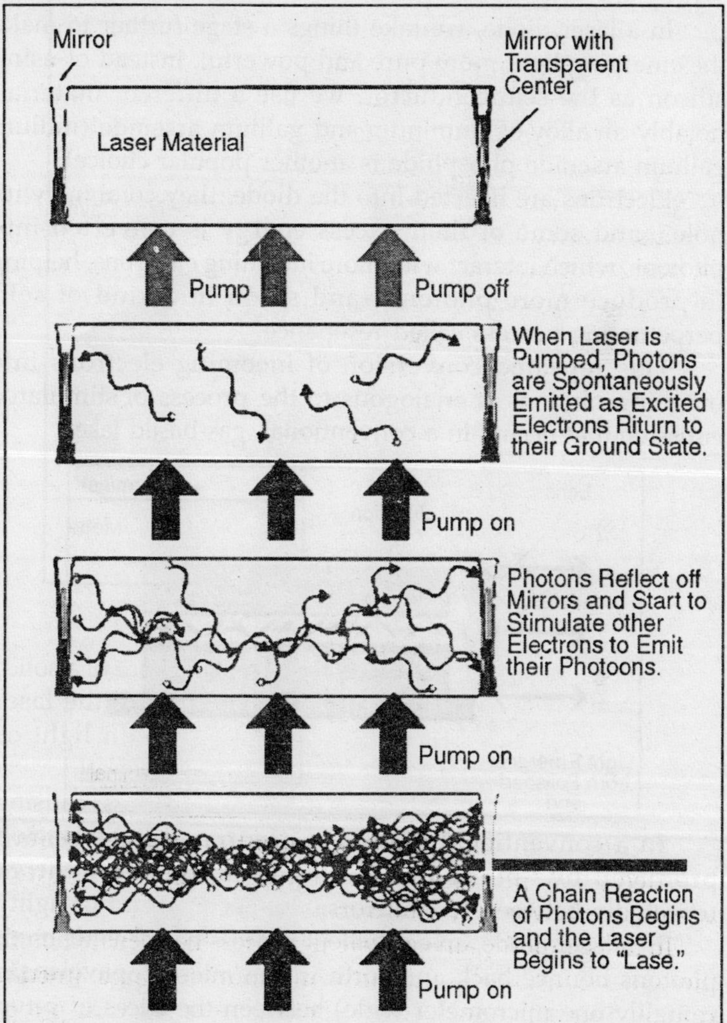

Laser Action

The combination of spontaneous emission first, and then

stimulated emission, causes the laser to "lase," which means it generates a coherent beam of light at a single frequency.

WORKING

In a laser diode, we take things a stage further to make the emerging light more pure and powerful. Instead of using silicon as the semiconductor, we use a different material, notably an alloy of aluminum and gallium arsenide (indium gallium arsenide phosphide is another popular choice).

Electrons are injected into the diode, they combine with holes, and some of their excess energy is converted into photons, which interact with more incoming electrons, helping to produce more photons—and so on in a kind of self-perpetuating process called resonance.

This repeated conversion of incoming electrons into outgoing photons is analogous to the process of stimulated emission that occurs in a conventional, gas-based laser.

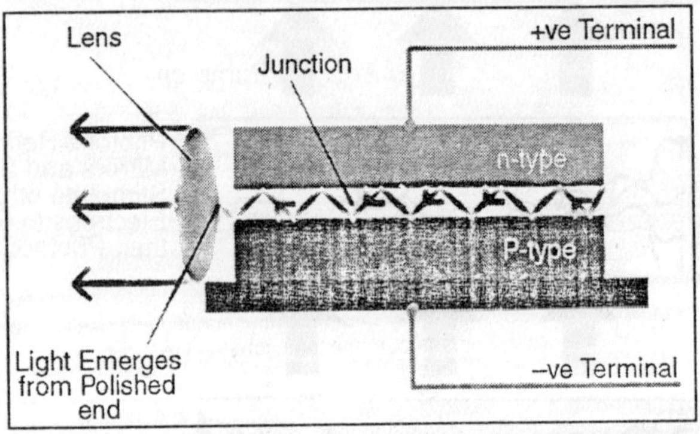

In a conventional laser, a concentrated light beam is produced by "pumping" the light emitted from atoms repeatedly between two mirrors.

In a laser diode, an equivalent process happens when the photons bounce back and forth in the microscopic junction (roughly one micrometer wide) between the slices of p-type and n-type semiconductor, which is technically known as a Fabry-Perot resonant cavity. The amplified laser light eventually emerges from the polished end of the gap in a beam parallel to the junction. From there, it goes on to read music from your CD, scan the price on your cornflakes, print out your college dissertation, or do a thousand other useful things!

TUNNEL DIODE

INTRODUCTION

The TUNNEL DIODE is a pn junction with a very high concentration of impurities in both the p and n regions. The high concentration of impurities causes it to exhibit the properties of a negative-resistance element over part of its range of operation, as shown in the characteristic curve in figure below:

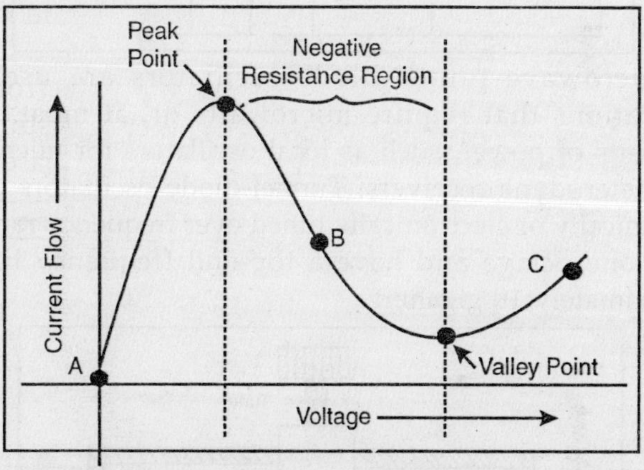

In other words, the resistance to current flow through the tunnel diode increases as the applied voltage increases over a portion of its region of operation. Outside the negative-resistance region, the tunnel diode functions essentially the same as a normal diode. However, the very high impurity density causes a junction depletion region so narrow that both holes and electrons can transfer across the pn junction by a quantum mechanical action called TUNNELING.

Tunneling causes the negative-resistance action and is so fast that no transit-time effects occur even at microwave frequencies. The lack of a transit-time effect permits the use of tunnel diodes in a wide variety of microwave circuits, such as amplifiers, oscillators, and switching devices.

TUNNEL-DIODE AS OSCILLATOR

A tunnel diode, biased at the centre point of the negative-resistance range (point B in figure above) and coupled to a tuned circuit or cavity, produces a very stable oscillator. The oscillation frequency is the same as the tuned circuit or cavity frequency.

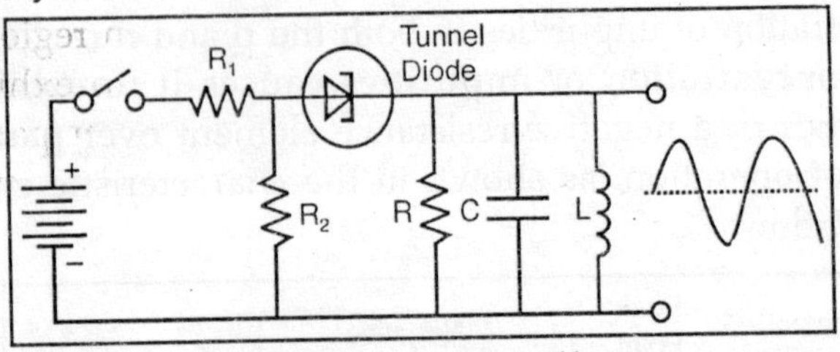

Microwave tunnel-diode oscillators are useful in applications that require microwatts or, at most, a few milliwatts of power, such as local oscillators for microwave superheterodyne receivers. Tunnel-diode oscillators can be mechanically or electronically tuned over frequency ranges of about one octave and have a top-end frequency limit of approximately 10 gigahertz.

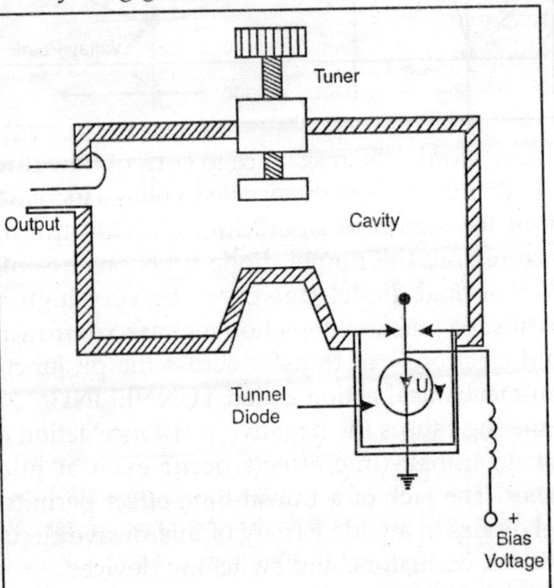

Tunnel-diode oscillators that are designed to operate at microwave frequencies generally use some form of transmission line as a tuned circuit. Suitable tuned circuits can be built from coaxial lines, transmission lines, and waveguides. An example of a highly stable tunnel-diode oscillator is shown in figure above.

A tunnel-diode is loosely coupled to a high-Q tunable cavity. Loose coupling is achieved by using a short, antenna feed probe placed off-centre in the cavity. Loose coupling is used to increase the stability of the oscillations and the output power over a wider bandwidth.

The output power produced is in the range of a few hundred microwatts, sufficient for many microwave applications. The frequency at which the oscillator operates is determined by the physical positioning of the tuner screw in the cavity. Changing the output frequency by this method is called MECHANICAL TUNING.

In addition to mechanical tuning, tunnel-diode oscillators may be tuned electronically. One method is called BIAS TUNING and involves nothing more than changing the bias voltage to change the bias point on the characteristic curve of the tunnel-diode. Another method is called VARACTOR TUNING and requires the addition of a varactor to the basic circuit.

TUNNEL-DIODE AS AMPLIFIERS

Low-noise, tunnel-diode amplifiers represent an important microwave application of tunnel diodes. Tunnel-diode amplifiers with frequencies up to 85 gigahertz have been built in waveguides, coaxial lines, and transmission lines. The low-noise generation, gain ratios of up to 30 dB, high reliability, and light weight make these amplifiers ideal for use as the first stage of amplification in communications and radar receivers.

Most microwave tunnel-diode amplifiers are reflection type,circulator-coupled amplifiers . As in oscillators, the tunnel diode is biased to the centre point of its negative-resistance region, but a CIRCULATOR replaces the tuned cavity.A circulator is a waveguide device that allows energy to travel in one direction only, as shown in figure below:

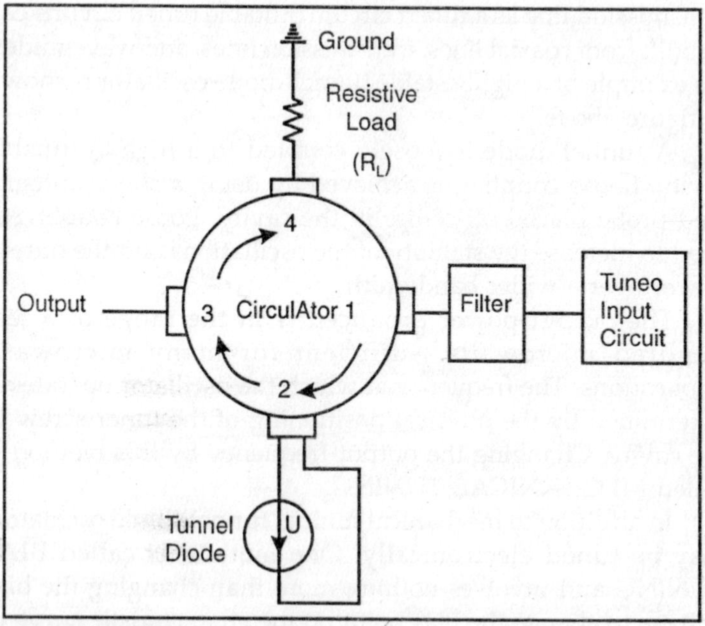

The tunnel diode in figure above is connected across a tuned-input circuit. This arrangement normally produces feedback that causes oscillations if the feedback is allowed to reflect back to the tuned-input circuit.

The feedback is prevented because the circulator carries all excess energy to the absorptive load (R_L). In this configuration the tunnel diode cannot oscillate, but will amplify.

The desired frequency input signal is fed to port 1 of the circulator through a bandpass filter. The filter serves a dual purpose as a bandwidth selector and an impedance-matching device that improves the gain of the amplifiers.

The input energy enters port 2 of the circulator and is amplified by the tunnel diode. The amplified energy is fed from port 2 to port 3 and on to the mixer. If any energy is reflected from port 3, it is passed to port 4, where it is absorbed by the matched load resistance.

TUNNEL-DIODE FREQUENCY CONVERTERS AND MIXERS

Tunnel diodes make excellent mixers and frequency converters because their current-voltage characteristics are highly nonlinear. While other types of frequency converters usually have a conversion power loss, tunnel-diode converters can actually have a conversion power gain.

A single tunnel diode can also be designed to act as both the nonlinear element in a converter and as the negative-resistance element in a local oscillator at the same time.

Practical tunnel-diode frequency converters usually have either a unity conversion gain or a small conversion loss. Conversion gains as high as 20 dB are possible if the tunnel diode is biased near or into the negative-resistance region. Although high gain is useful in some applications, it presents problems in stability.

For example, the greatly increased sensitivity to variations in input impedance can cause high-gain converters to be unstable unless they are protected by isolation circuitry.

As with tunnel-diode amplifiers, low-noise generation is one of the more attractive characteristics of tunnel-diode frequency converters.

Low-noise generation is a primary concern in the design of today's extremely sensitive communications and radar receivers. This is one reason tunnel-diode circuits are finding increasingly wide application in these fields.

IMPATT DIODE

INTRODUCTION

A semiconductor microwave diode that, when its junction is biased into avalanche, exhibits a negative resistance over a frequency range determined by the transit time of charge carriers through the depletion region. IMPact Avalanche Transit Time diode is a high power radio frequency (RF) generator operating from 3 to 100 gHz. IMPATT diodes are fabricated from silicon, gallium arsenide, or silicon carbide.

An IMPATT diode is reverse biased above the breakdown voltage. The high doping levels produce a thin depletion region. The resulting high electric field rapidly accelerates carriers which free other carriers in collisions with the crystal lattice. Holes are swept into the P_+ region. Electrons drift toward the N regions.

The cascading effect creates an avalanche current which increases even as voltage across the junction decreases. The pulses of current lag the voltage peak across the junction. A "negative resistance" effect in conjunction with a resonant circuit produces oscillations at high power levels (high for semiconductors).

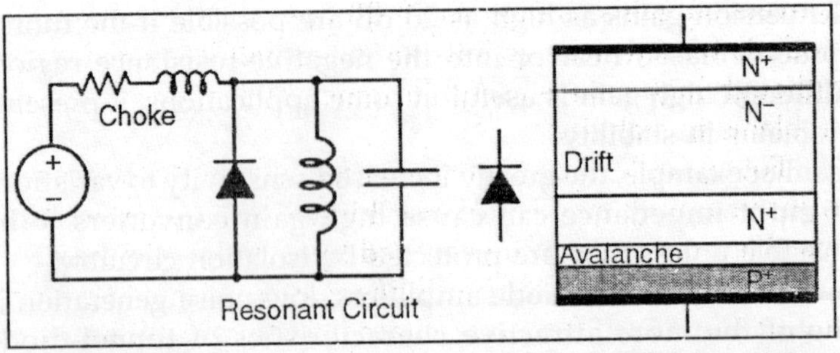

IMPATT diode: Oscillator circuit and heavily doped P and N layers.

The resonant circuit in the schematic diagram of Figure is the lumped circuit equivalent of a waveguide section, where the IMPATT diode is mounted. DC reverse bias is applied through a choke which keeps RF from being lost in the bias supply. This may be a section of waveguide known as a bias Tee. Low power RADAR transmitters may use an IMPATT diode as a power source. They are too noisy for use in the receiver.

THE GUNN DIODE

INTRODUCTION

A Gunn diode, also known as a transferred electron device (TED), is a form of diode used in high-frequency electronics. It is somewhat unusual in that it consists only of doped semiconductor material, whereas most diodes consist of both P and N-doped regions. In the Gunn diode, three regions exist: two of them are heavily N-doped on each terminal, with a thin layer of lightly doped material in between.

When a voltage is applied to the device, the electrical gradient will be largest across the thin middle layer. Conduction will take place as in any conductive material with current being proportional to the applied voltage.

Eventually, at higher field values, the conductive properties of the middle layer will be altered, increasing its resistivity and reducing the gradient across it, preventing further conduction and current actually starts to fall down. In practice, this means a Gunn diode has a region of negative differential resistance.

A gunn diode is solely composed of N-type semiconductor. As such, it is not a true diode. Figure shows a lightly doped N_- layer surrounded by heavily doped N+ layers. A voltage applied across the N-type gallium arsenide gunn diode creates a strong electric field across the lightly doped N^- layer.

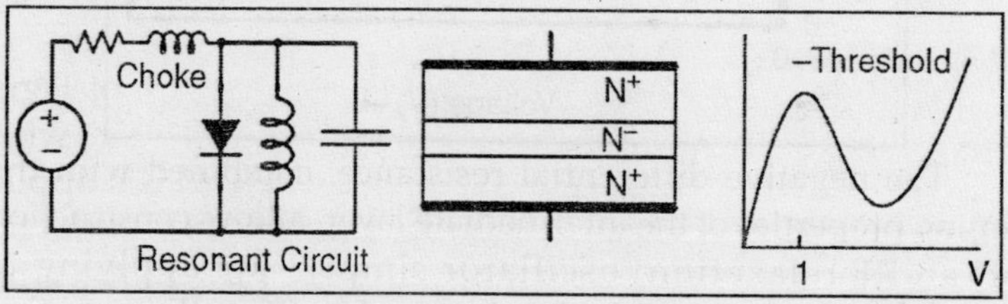

Gunn diode: Oscillator circuit and cross section of only N-type semiconductor diode.

As voltage is increased, conduction increases due to electrons in a low energy conduction band. As voltage is increased beyond the threshold of approximately 1 V, electrons move from the lower conduction band to the higher energy conduction band where they no longer contribute to conduction. In other words, as voltage increases, current decreases, a negative resistance condition. The oscillation frequency is determined by the transit time of the conduction electrons, which is inversely related to the thickness of the N^- layer. The frequency may be controlled to some extent by embedding the gunn diode into a resonant circuit. The lumped circuit equivalent shown in Figure is actually a coaxial transmission line or waveguide. Gallium arsenide gunn diodes are available for operation from 10 to 200 gHz at 5 to 65 mw power. Gunn diodes may also serve as amplifiers.

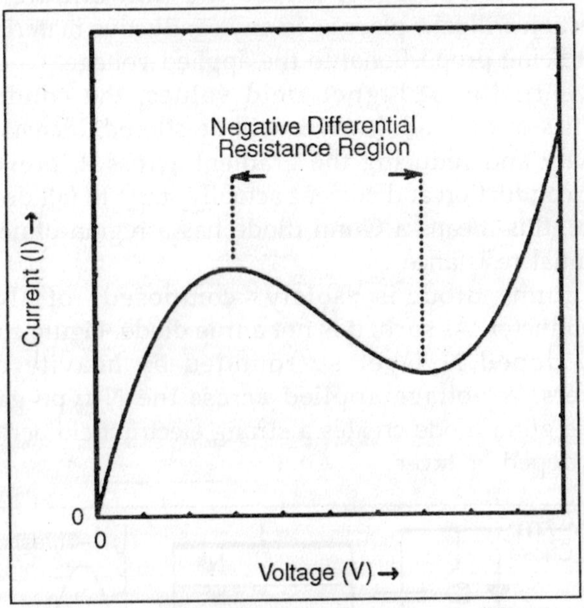

The negative differential resistance, combined with the timing properties of the intermediate layer, allows construction of an RF relaxation oscillator simply by applying a suitable current through the device. In effect, the negative differential resistance created by the diode will negate the real and positive resistance of an actual load and thus create a "zero" resistance circuit which will sustain oscillations indefinitely.

The oscillation frequency is determined partly by the properties of the thin middle layer, but can be tuned by external factors. Gunn diodes are therefore used to build oscillators in the 10 GHz and higher (THz) frequency range, where a resonator is usually added to control frequency. This resonator can be take the form of a wave guide, microwave cavity or YIG sphere. Tuning is done mechanically, by adjusting the parameters of the resonator, or in case of YIG spheres by changing the magnetic field.

P-N-P-N DIODE

The four-layer diode, also called the Shockley diode after its inventor William Shockley, is essentially a low-current SCR without a gate. It is classified as a diode because it has only two external terminals through anode and cathode. Because of its four doped regions it is often called a P-N-P-N diode. The basic structure, two transistor version, equivalent transistor circuit and schematic symbol of a Shockley diode are shown in figure.

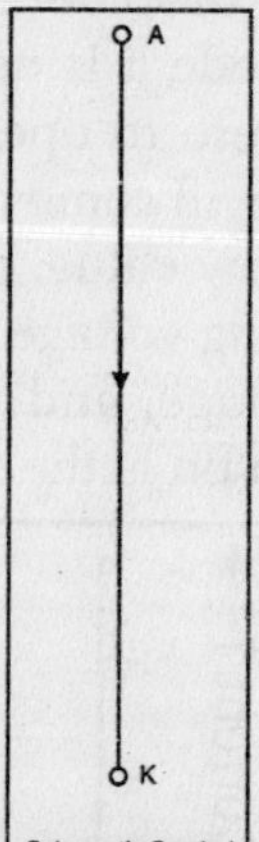

Schematic Symbol

The easiest way to understand how it operates is to visualize it to be formed of two transistors Q_x and Q_2 placed back to back, as shown in figure.b. Figure shows the diode split-up into two parts, displaced mechanically but connected electrically.

The left half is a P-N-P transistor while the right half is an N-PN- transistor. It may be seen that N-type base region of P-N-P transistor forms the collector of N-P-N transistor while P-type base of N-P-N transistor forms the collector of P-N-P transistor. Thus the four-layer diode is equivalent to the latch shown in figure.

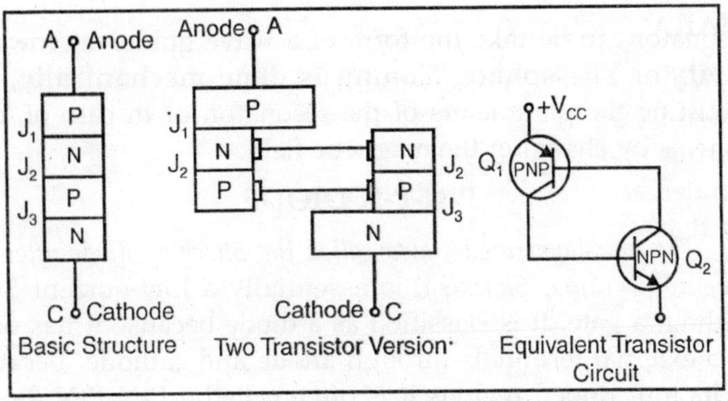

Because there are no trigger inputs, the only way to switch the device on is to increase the anode-to-cathode voltage V_{AK} to the forward switching voltage, and the only way to open it is by low current drop out.

With a four layer diode it is not necessary to reduce the current all the way to zero to open the latch. The inter nal transistors of the device will come out of saturation when the current is reduced to a low value, called the holding current.

The forward switching voltage V_s is equivalent of the SCR for ward breakover voltage, and the minimum cur rent at which device will switch on is the switch-ing current I_S.

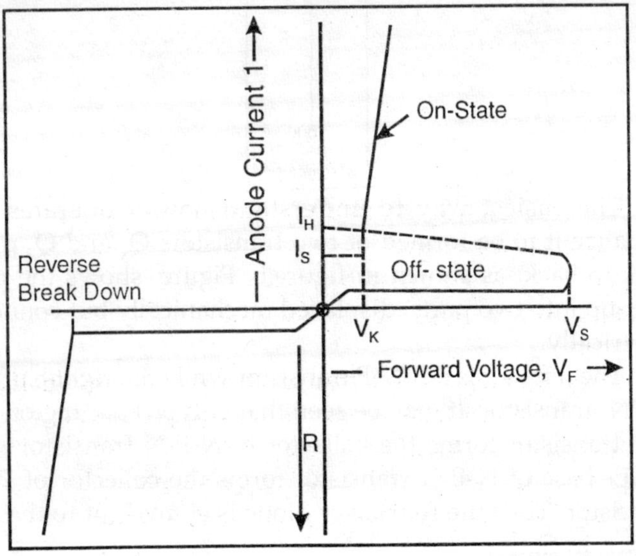

CHARACTERISTICS

Voltage-current characteristic of a Shockley diode is shown in figure. The device has two operating states: con ducting and non-conducting. In non-conducting state, it operates on lower line with negligible current and a voltage less than switching voltage or breakover voltage. When the voltage tries to ex ceed the breakover voltage, the device breaks down and switches along the dotted line to the conducting or on-state.

The dotted line indicates an unstable or a temporary condition. The device can have current and voltage values on this dotted line only briefly as it switches between the two stable operating states. In conducting state or in on-state, the device operates on the upper line. As long as the current through the device is greater than the holding current I_H, then the voltage across it is slightly greater than knee voltage, V_K. When the current falls below the level of the holding current I_H, the device switches back along the dotted line to the non-conducting or off-state.

APPLICATION

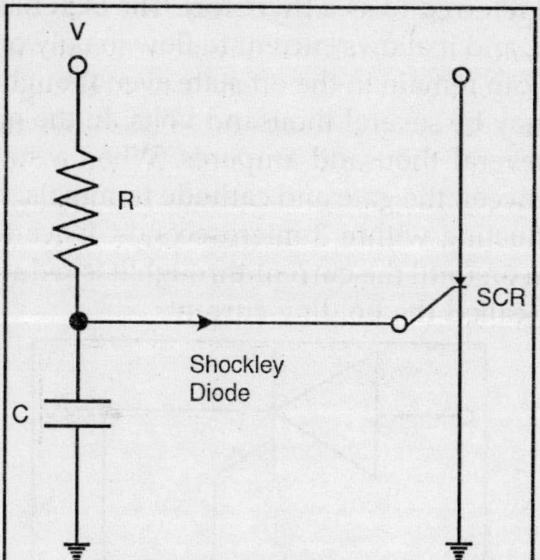

Shockley Diode As a Trigger Swilch

One common application of the Shockley diode is as a *trigger switch for an SCR*. The circuit is shown in figure. When the circuit is energized, the capacitor will start getting charged and eventually, the voltage across the capacitor will be sufficiently high to first turn-on Shockley diode and then the SCR.Another application of this diode is as a relaxation oscillator.

SEMICONDUCTOR CONTROLLED RECTIFIER (SCR)

INTRODUCTION

A silicon-controlled rectifier (or semiconductor-controlled rectifier) is a 4-layersolid-state device that controls current flow. The name "silicon controlled rectifier" is a trade name for the type of thyristor commercialized at General Electric.

An SCR can be seen as a conventional rectifier controlled by a gate signal. It is a 4-layered 3-terminal device. A p-type layer acts as anode and an n-type layer as a cathode; the p-type layer closer to the n-type acts as a gate. The heart of the SCR power controller is the SCR (silicon controlled rectifier, also sometimes referred to as a thyristor). The SCR has two states, ON or OFF, and it allows current to flow in only one direction.

SCR's can remain in the off state even though the applied potential may be several thousand volts. In the on state, they can pass several thousand amperes. When a small signal is applied between the gate and cathode terminals, the SCR will begin conducting within 3 microseconds. Once turned on, it will remain on until the current through it is reduced to a very low value, called the holding current.

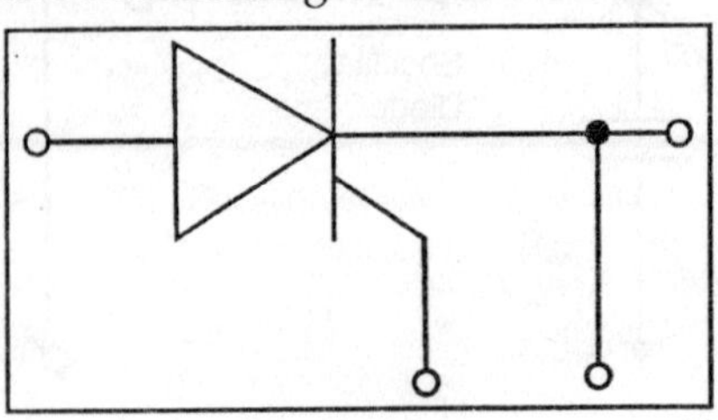

Fig. SCR Symbol

Because the SCR allows current to flow in only one direction, two SCR's are connected in an inverse parallel configuration to control AC current.

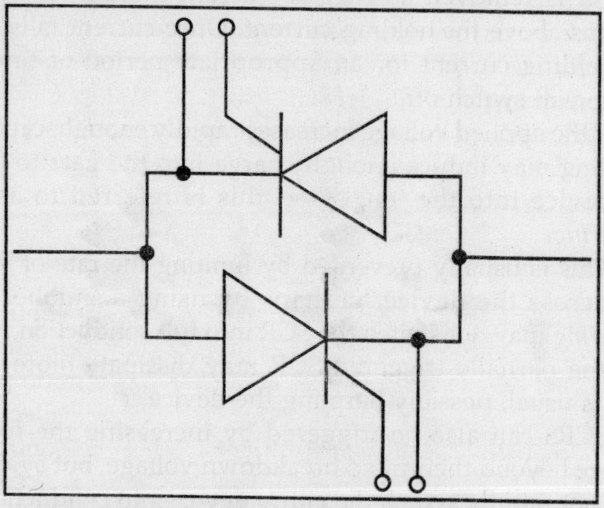

Fig. AC "Back to Back" Switch

Three SCR construction styles are available: the disc (a.k.a. hockey puck), the module, and the stud mount. Modern SCR controllers generally use hockey puck or module construction.

Hockey puck: The hockey puck style is an assembly that has essentially the same physical shape as a hockey puck. The construction provides excellent cooling of the semiconductor material and is generally used in higher current applications.

Module: SCR modules are assemblies in which the SCR's are contained in a plastic enclosure with an electrically isolated mounting plate. SCR modules are becoming increasingly popular and modules containing a variety of SCR configurations are available. They are easy and inexpensive to mount to a heat sink, they typically have large surge current ratings and they provide electrical isolation, allowing multiple modules to be mounted on a common heat sink.

MODES OF OPERATION

In the normal 'off' state the device restricts current flow

to the leakage current. When the gate to cathode voltage exceeds a certain threshold, the device turns 'on' and conducts current. The device will remain in the 'on' state even after gate current is removed so long as current through the device remains above the holding current. Once current falls below the holding current for an appropriate period of time, the device will switch off.

If the applied voltage increases rapidly enough, capacitive coupling may induce enough charge into the gate to trigger the device into the 'on' state; this is referred to as *dv/dt triggering*.

This is usually prevented by limiting the rate of voltage rise across the device, perhaps by using a snubber. *dv/dt triggering* may not switch the SCR into full conduction rapidly and the partially-triggered SCR may dissipate more power than is usual, possibly harming the device.

SCRs can also be triggered by increasing the forward voltage beyond their rated breakdown voltage, but again, this does not rapidly switch the entire device into conduction and so may be harmful so this mode of operation is also usually avoided. Also, the actual breakdown voltage may be substantially higher than the rated breakdown voltage, so the exact trigger point will vary from device to device.

Control Modes

The power delivered to a load may be regulated or proportioned by SCR power controllers using either the phase-angle or the integral cycle (zero-cross voltage switching) control mode. Each control mode has its own specific advantages and disadvantages and each application should be reviewed to determine the most compatible mode of control.

Phase-angle

In phase-angle control each SCR of the back-to-back pair is turned on for a variable portion of the half-cycle that it conducts. Power is regulated by advancing or delaying the point at which the SCR is turned ON within each half cycle. Light dimmers are an example of phase-angle control.

Phase-angle control provides a very fine resolution of power and is used to control fast responding loads such as tungsten-filament lamps or loads in which the resistance changes as a function of temperature. Phase-angle control is required if the load is transformer-coupled or inductive.

Phase-angle controllers are typically more expensive than zero-cross controllers because the phase-angle circuit requires more sophistication than does a zero-cross circuit. Phase-angle control of three-phase power requires SCR's in all three legs and is appreciably more expensive than zero-cross control which only requires SCR's in two of the three legs.

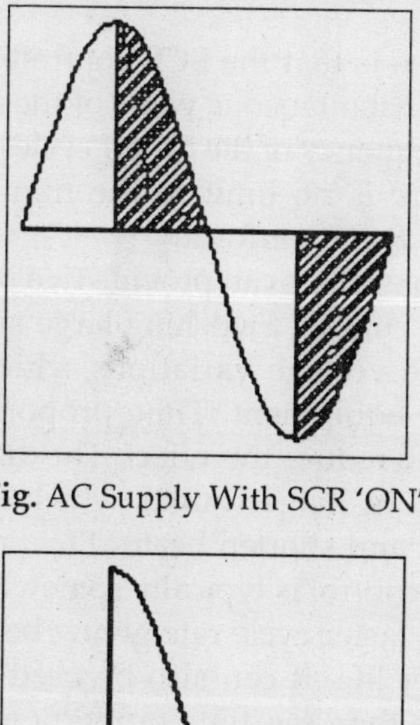

Fig. AC Supply With SCR 'ON'

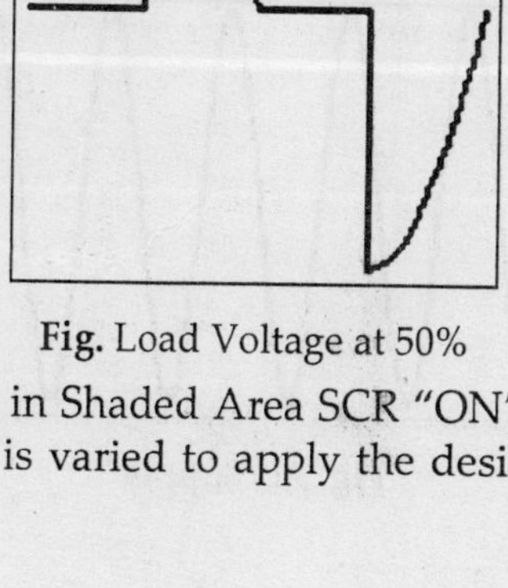

Fig. Load Voltage at 50%

Time Shown in Shaded Area SCR "ON" time, shown by the shaded area, is varied to apply the desired load voltage.

Power is regulated by advancing or delaying the point at which the SCR's are turned on.

Zero-cross

The term zero-cross or synchronous operation of SCR's is derived from the fact that the SCR's are turned on only when the instantaneous value of the sinusoidal waveform is zero. In zero-cross operation power is applied for a number of continuous half-cycles and then removed for a number of half-cycles to achieve the desired load power in the same manner as power would be controlled with a mechanical switching device.

The difference is that the SCR controllers always switch power when the instantaneous value of the applied voltage is zero. Also, the frequency of the on-off cycles can be extremely fast because there is no limit to the number of switching operations the SCR can perform.

Zero-cross controllers can provide two rather distinctively different types of control. Switching large amounts of current can result in line voltage variations which affect ambient lighting or other equipment. Time proportioning control is sometimes used to reduce this effect. The disadvantage is that power is applied in longer bursts which can in turn cause control problems and shorten heater life.

Distributive control is typically somewhat less expensive, provides a much faster cycle rate giving better controllability and longer heater life. It can also be used with much faster responding loads than can time proportioning.

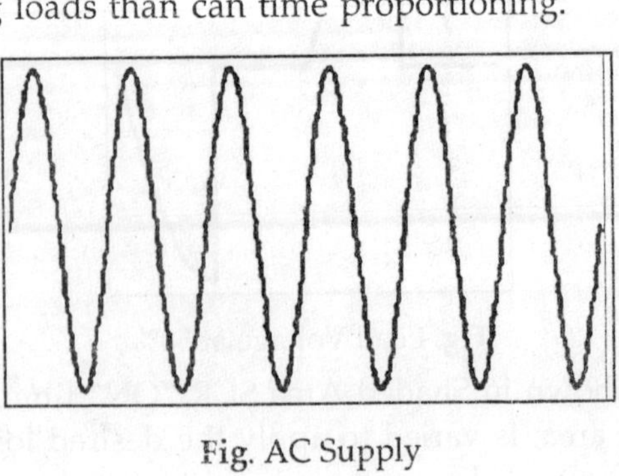

Fig. AC Supply

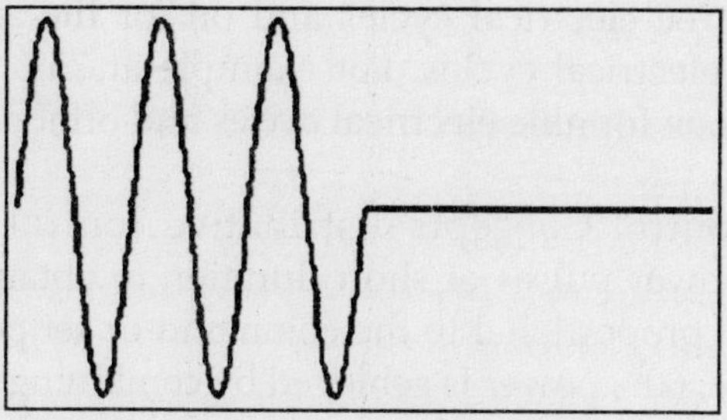

Fig. Zero Cross at 50%

Zero-cross Time Proportioning

Zero-cross time proportion control is accomplished with a fixed or constant time base, therefore the total of power "ON" time and power "OFF" time is always equal to a fixed value. For example, if the time base is ten seconds and the desired power is 50%, then power is applied for 5 seconds and removed for 5 seconds.

If the desired power were 25% then power is applied for 2.5 seconds and removed for the remaining period of 7.5 seconds.

The disadvantage of time proportioning particularly as the time base is increased is that the load temperature varies considerably between the on-off cycles. This can shorten the life of heater elements and decrease the ability to obtain precise process control.

Distributive Zero-cross

Distributive zero-cross does not use a fixed or constant time base as is used in time proportioning. The technique used by Control Concepts applies load power for 3 electrical cycles and removes load power for 3 electrical cycles at 50% power. At lower power requirements the controller will apply power for 3 electrical cycles and is then off for the appropriate number of electrical cycles.

For example, at 25% power the controller is on for 3 electrical cycles and off for 9 electrical cycles, or on for 3 electrical cycles out of 12. At higher power levels the controller is off for three electrical cycles and on for the appropriate number of electrical cycles. For example at 75% power the controller is on for nine electrical cycles and off for 3 electrical cycles.

The Control Concepts distributive control technique combines power pulses of short duration to obtain the exact power level proportional to the command or set point signal. For example, 60% power is achieved by combining two power pulses. The first power pulse consists of 4 on cycles and 3 off cycles; the second power pulse consists of 5 on cycles and 3 off cycles, providing a total of 9 on cycles during 15 cycles.

Of course, such rapid, short bursts of power would be impossible with mechanical contactors. Zero-cross control is typically less expensive than phase-angle control and generates fewer harmonics.

However, zero-cross can only be used to control power to resistive loads that do not change appreciably with temperature or time and which are directly-coupled (no transformer between the SCR controller and the load).

Zero-cross control of large power levels can cause supply voltage fluctuations resulting in ambient lighting fluctuations or other problems and can cause overheating of transformers supplying power to the controller and load.

Power Factors

Power Factor and Harmonics

The operation of zero-cross or phase-angle SCR controllers can lower the power factor, resulting in a higher electrical cost. Harmonics can be generated which may cause radio frequency interference (RFI) or possibly affect the operation of other equipment.

Power Factor

The power factor of a single-phase circuit is the ratio of true power in watts, as measured with a wattmeter, to the apparent power in volt-amperes, obtained as the product of voltage and current.

P.F. = KW/KVA

The power factor of a balanced polyphase circuit is the same as that of the single phases. When the phases are not balanced the true power factor is indeterminate.)

Consider the supply voltage ES to be 480 volts and the load voltage EL controlled by the SCR controller to be 240 volts.

The load current IL is, of course, equal to the supply current IS and is equal to EL/RL or 100 amps. The KW of the system is 24 (240 volts × 100 amps) and the KVA is of the system is 48 (480 volts × 100 amps).

P.F. = 24/48 = EL × IL/ES × IL = EL/ES = (% Applied power/100)0.5 = 0.5

The power factor with resistive loads controlled by either phase-angle or zero-cross control is the ratio of the load voltage to the supply voltage and is therefore proportional to (% applied power/100)0.5.

Tests have shown that most power factor meters supplied by utility companies respond correctly to phase-angle control, but provide near unity power factor for zero-cross control.

Newer meters using solid state technology and which sample for a longer time period provide power factor measurements very close to the theoretical values for both phase-angle and zero-cross control.

Harmonics and RFI

Harmonic wave forms are generated when electrical power is switched and are therefore generated when power is controlled by mechanical contacts, saturable core reactors, SCR controllers and all other power switching devices.

Harmonics are sinusoidal waveforms with frequencies that are integral multiples of the fundamental frequency. For example, a waveform that has twice the frequency of the fundamental frequency is called the second harmonic.

All repetitive waveforms are formed by the addition of harmonics. In general, abrupt changes in the waveform or complex waveforms cause the magnitudes of the harmonics to increase, increasing the possibility of interference problems.

In SCR control the magnitude of the harmonics are the greatest when the load power is controlled at 50%. The table below shows the relative magnitude and frequency for zero-cross and phase-angle control.

Although harmonics and RFI are a common concern, very few if any problems have occurred in actual application. This is because harmonics are attenuated by inductance, and because the effects of harmonics are attenuated by proper shielding and grounding of electrical equipment.

Comparison of Harmonics in Phase-Angle and Zero-Cross Control

Zero-Cross			Phase-Angle		
Harmonic	Frequency (Hz)	Magnitude (%)	Harmonic	Frequency (Hz)	Magnitude (%)
1	10	10.90	1	60	59.70
2	20	0.00	2	120	0.00
3	30	14.10	3	180	31.80
4	40	0.00	4	240	0.00
5	50	34.71	5	300	10.60
6	60	50.00	6	360	0.00
7	70	29.80	7	420	10.60
8	80	0.00	8	480	0.00
9	90	8.49	9	540	6.40
10	100	0.00	10	600	0.00
11	110	4.50	11	660	6.40
12	120	0.00	12	720	0.00
13	130	2.88	13	780	4.60
14	140	0.00	14	840	0.00
15	150	2.03	15	900	4.60
16	160	0.00	16	960	0.00
17	170	1.50	17	1020	3.50
18	180	0.00	18	1080	0.00
19	190	1.18	19	1140	3.50
20	200	0.00	20	1200	0.00
21	210	0.09	21	1260	2.90
22	220	0.00	22	1320	0.00

Table: Comparison of Harmonics in Phase-Angle and Zero-Cross Control

Zero-Cross Harmonic (Hz)	Frequency	Magnitude (%)	Phase-Angle Harmonic (Hz)	Frequency (%)	Magnitude
11	110	4.50	11	660	6.40
12	120	0.00	12	720	0.00
13	130	2.88	13	780	4.60
14	140	0.00	14	840	0.00
15	150	2.03	15	900	4.60
16	160	0.00	16	960	0.00
17	170	1.50	17	1020	3.50
18	180	0.00	18	1080	0.00
19	190	1.18	19	1140	3.50
20	200	0.00	20	1200	0.00
21	210	0.09	21	1260	2.90
22	220	0.00	22	1320	0.00

PERFORMANCE AND FEATURES

Soft Start & Missing Cycle Detection

These features are required in phase-angle control to assure that the load power is gradually increased from zero to the value set by the command signal if load power is interrupted. This gradual increase of power prevents surge currents and avoids the possibility of saturating inductive loads like transformers.

Current Limiting

This feature, available only on phase-angle controllers, provides a means to prevent the ioad current from exceeding a preset value.

Current limiting is used to protect the load, the SCR controller, fuses, and the system supply from large surge currents that could occur at start up due to loads that have a low resistance when cold.

Feedback/Feed Forward

Either feedback or feed forward provides the means to achieve a linear relation between the desired output and the control or command signal. Feedback implies that the desired parameter is measured and "fed back" to the input of the control such that the output can be increased or decreased if corrective action is required.

Feed forward involves a less expensive technique in which the output is simulated in the circuitry and corrective action is taken based upon the simulated circuit response. Both techniques provide a linear output with respect to the control or command signals and eliminate the effects of load and supply variations.

Over Current Trip

The over current trip feature prevents the SCR's from being turned on if the SCR current has exceeded a preset value during the last half cycle.

Essentially, this feature is an electronic fuse which eliminates the need for expensive and troublesome semi-conductor fuses.

The controller can be reset by either removing and re-applying power or by momentarily closing a remote switch. This feature has proven to be a more reliable protection technique than fast acting fuses.

Shorted SCR Detection

If an SCR does fail, it typically becomes shorted, allowing load current to flow continuously. In the event an SCR shorts, this feature energizes a relay which can be used to activate an alarm or remove system power.

Sync-Guard

This feature, available on some zero-cross controllers produced by Control Concepts, reduces the possibility of synchronous operation of two or more controllers. It reduces the variations in power demand resulting in a more stable supply voltage and improved power factor.

The Sync-Guard feature does not alter the power applied to the load, but adjusts the time when power is applied in such a manner as to reduce the possibility of the controllers being ON and OFF at the same time.

Trans-Guard

The use of zero-cross controllers on the secondary of a transformer can cause saturation of the transformer resulting in excess transformer temperatures and early failure.

Transformers can be caused to saturate if a DC voltage is applied to the primary. DC voltage can be induced on the primary by DC components in the secondary.

A simple half wave rectifier circuit will induce a DC voltage on the primary due to the fact that the voltage drop across the source resistance during the half cycle the diodeconducts, will lower the primary voltage when the diode conducts.

The effect is the same as if a zero-cross controller were not providing an equal number of positive and negative half cycles to the load. The source inductance can also cause a DC voltage to occur on the transformer primary.

The Trans-Guard feature eliminates the problem by always supplying an odd number of ON half-cycles and an even number of OFF half-cycles. This technique guarantees that no DC will occur independent of the source impedance or the load configuration.

The saturation problem is not likely to occur when the load is a small percentage of the transformer capacity, or when the source resistance and inductance are small. However, the potential problem always exists unless the controller is designed with the feature described above.

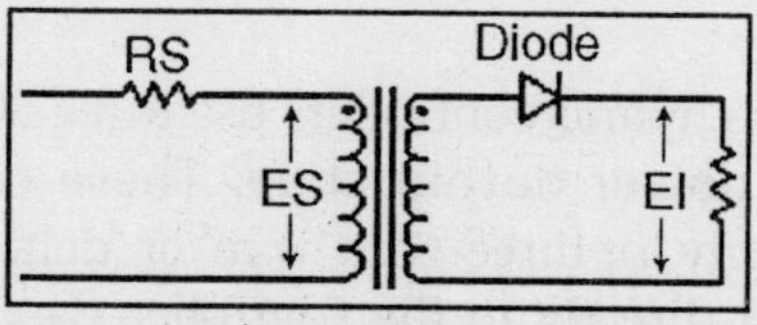

Fig. Transformer Saturation Due to Secondary DC Current

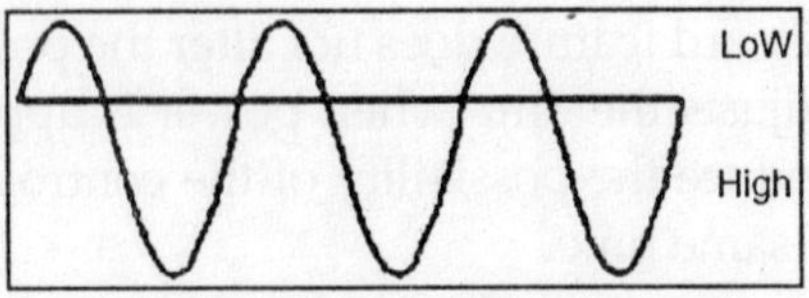

Fig. Transformer Voltage ES

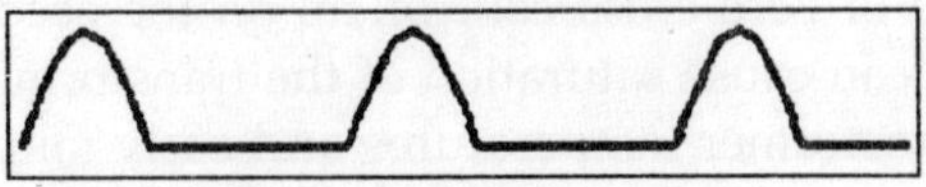

Fig. Load Voltage EL

Load

Load configurations and considerations for single-phase control should be obvious from the previous discussion. However, there are several unique load configurations and some important considerations which need to be examined in three-phase applications.

Single-Phase Load Connections

The connections for single-phase are the same for phase-angle and zero-cross. The difference between the two being the manner in which the SCR's are controlled.

Three-Phase Load Connections

Phase-Angle Control

Phase-angle control of three-phase requires a total of 6 power switching devices. These devices can be configured as hybrid (3 SCR's & 3 diodes), 6 SCR in-line or 6 SCR inside delta.

Hybrid Control

Three-phase hybrid controllers use three SCR's and three diodes in the power output state. These controllers are intended primarily for three-wire 'wye' or 'delta' type resistive loads connected directly to the controller (i.e., a transformer is not connected between the load and the controller). The three-phase hybrid controller should not be used in a four wire circuit. The fourth wire common return would allow uncontrolled conduction through the power diodes resulting in approximately 50% power output even though the SCR's were off.

The advantage of hybrid controllers is that the cost is somewhat less than six SCR control because the circuit is less complex and diodes are less costly. However, hybrid controllers should not be used if the load is unbalanced. An unbalanced load controlled with a hybrid controller will result in a DC current flow which can result in supply transformer saturation.

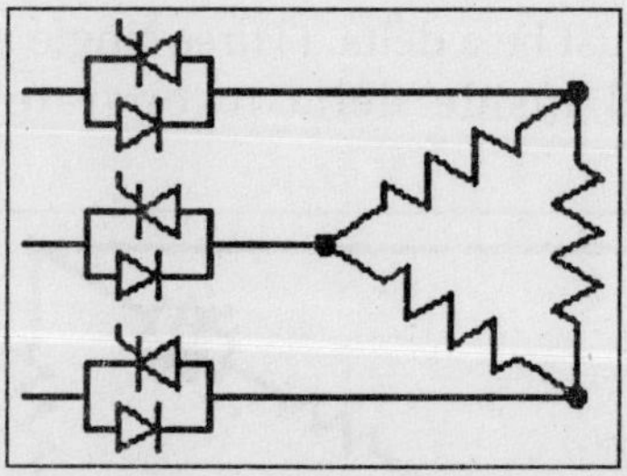

Fig. Hybrid Control: 3 SCR's & 3 Diodes

Six SCR in-line controllers

The Six SCR controller, as the name implies, uses six SCR's to control the load power. This configuration is an ideal configuration for inductive load applications, unbalanced resistive loads and transformer coupled loads. Caution: if the controller is operating a transformer, either the primary or the secondary must be a delta configuration.

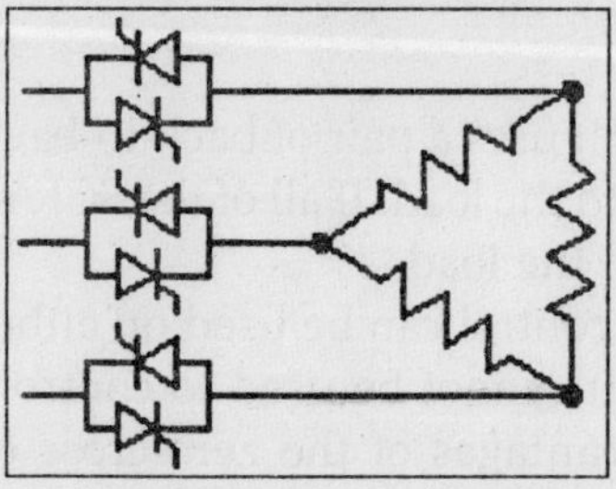

Fig. Six SCR In-Line Control

Six SCR inside delta control

Three phase inside delta control is essentially three single-phase controllers operating from the same command or set point.

Typically a six SCR controller can be configured to operate inside delta. It is also possible, of course, to configure three single phase controllers controlled by the same command signal. Because the current in each phase is less 57.74% of the line current, smaller and less expensive controllers can be used for inside delta control.

Caution: if inside delta control is used in the primary of a three-phase transformer, a six SCR controller must be used and the secondary must be a delta. (Three single phase controllers cannot be used inside delta to operate a three-phase transformer).

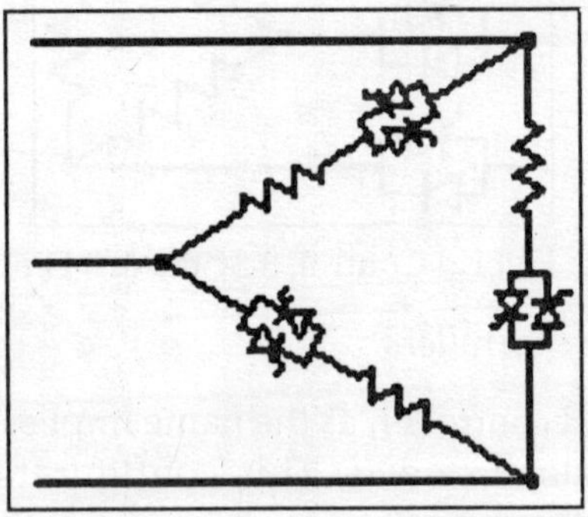

Fig. Six SCR Inside Delta Control

Zero-Cross Control

Two leg control

This approach uses a pair of back-to-back SCR's in two of the supply leads to the load. If all of the SCR's are off no power can be applied to the load.

The two leg control can be used on either 'wye' or 'delta' loads, however, it cannot be used to control four wire loads. The primary advantages of the zero-cross control technique are lower cost and extremely low RFI.

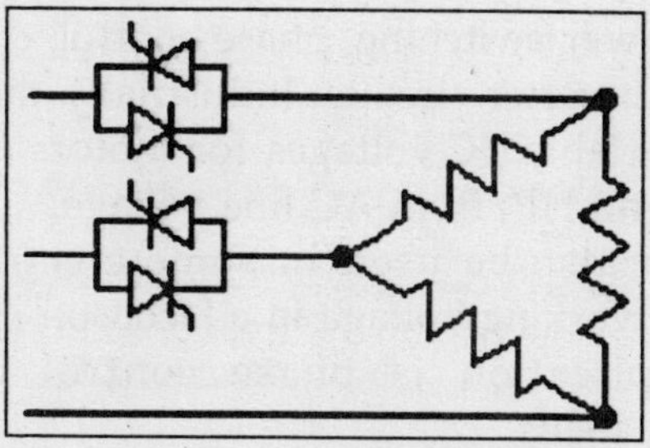

Fig.Two-Leg Zero-Cross

Three leg control

Some installations require that all three legs of the load be controlled and therefore back-to-back SCR's are placed in all three supply leads. The cost advantage of zero-cross control is reduced with 3-leg control because of the added SCR's and heat sink costs, however, the circuit cost is less than an equivalent phase-angle controller and the advantage of low RFI is retained.

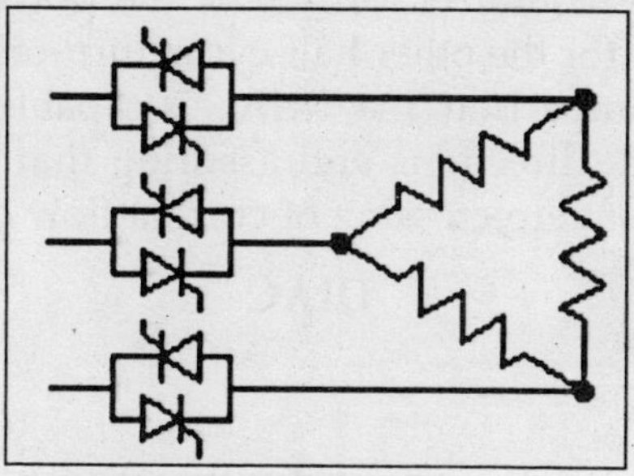

Fig. Three-Leg Zero-Cross

Inside delta

Zero-cross using the inside delta technique can be used to reduce the current rating of the SCR's, and in some instances, system cost can be reduced.

APPLICATIONS

SCRs are made with voltage ratings of up to 2500 volts, and with current ratings up to 3000 amperes per device. SCRs are used in power switching, phase control, chopper, battery chargers, and inverter circuits. Industrially they are applied to produce variable DC voltages for motors (from a few to several thousand HP) from AC line voltage.

They can also be used in some electric vehicles to modulate the working voltage in a Jacobson circuit. Another common application is phase control circuits used with inductive loads.

SCRs can also be found in welding power supplies where they are used to maintain a constant output current or voltage. Large silicon-controlled rectifer assemblies with many individual devices connected in series are used in high-voltage DC converter stations.

Much smaller SCRs are also used as ESD (electrostatic discharge) protection circuits on modern CMOS integrated circuits (see literature).

Two SCRs in *inverse parallel* are often used in place of a TRIAC for switching inductive loads on AC circuits. Because each SCR only conducts for half of the power cycle and is reverse-biased for the other half-cycle, turn-off of the SCRs is assured. By comparison, the TRIAC is capable of conducting current in both directions and assuring that it switches off during the brief zero-crossing of current flow can be difficult.

DIAC

DEFINITION

A diac is a full-wave or bi-directional semiconductor switch that can be turned on in both forward and reverse polarities. The name diac comes from the words Diode AC switch.

The diac is an electronics component that is widely used to assist even triggering of a triac when used in AC switches and as a result they are often found in light dimmers such as those used in domestic lighting.

These electronic components are also widely used in starter circuits for fluorescent lamps. Although the term is not often seen, they may also be called symmetrical trigger diodes

- a term resulting from the symmetry of their characteristic curve. The diac symbol used to depict this electronic component in circuit diagrams can be remembered as a combination of what may appear to be two diodes in parallel with each other but connected in opposite directions.

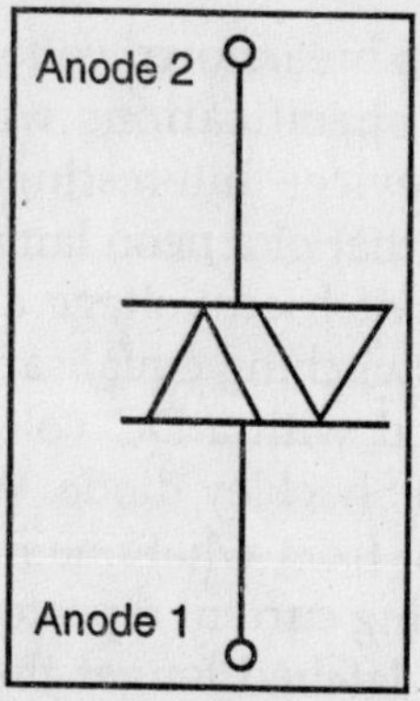

Fig. Circuit Symbol for the Diac

Owing to the fact that diacs are bi-direction devices the terminals cannot be labelled as anode and cathode as they are for a diode. Instead they may be labelled as A1 and A2 or MT1 ("Main Terminal") and MT2.

OPERATION

Diac circuits use the fact that a diac only conducts current only after a certain breakdown voltage has been exceeded. The actual breakdown voltage will depend upon the specification for the particular component type.

When the diac breakdown voltage occurs, the resistance of the component decreases abruptly and this leads to a sharp decrease in the voltage drop across the diac, and a corresponding increase in current. The diac will remain in its conducing state until the current flow through it drops below a particular value known as the holding current. When the current falls below the holding current, the diac switches back to its high resistance, or non-conducting state.

Diacs are widely used in AC applications and it is found that the device is "reset" to its non-conducting state, each time the voltage on the cycle falls so that the current falls below the holding current. As the behaviour of the device is approximately equal in both directions, it can provide a method of providing equal switching for both halves of an AC cycle, e.g for triacs.

Most diacs have a breakdown voltage of around 30 volts, although the exact specifications will depend upon the particular type of device.. Interestingly their behaviour is somewhat similar to that of a neon lamp, although they offer a far more precise switch on voltage and thereby provide a far better degree of switching equalisation.

A DIAC operated with a DC voltage across it behaves exactly the same as a Shockley diode. With AC, however, the behaviour is different from what one might expect.

Because alternating current repeatedly reverses direction, DIACs will not stay latched longer than one-half cycle. If a DIAC becomes latched, it will continue to conduct current only as long as voltage is available to push enough current in that direction.

When the AC polarity reverses, as it must twice per cycle, the DIAC will drop out due to insufficient current, necessitating another breakover before it conducts again. The result is the current waveform in Figure below.

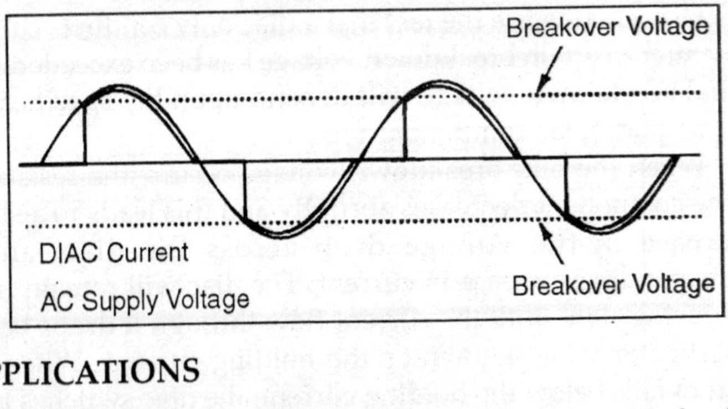

APPLICATIONS

One of the major uses of diacs within triac circuits. The diac is placed in series with the gate of a triac to provide a more symmetrical switching characteristic. It is found that triacs do not fire symmetrically as a result of slight differences between the two halves of the device. This results in harmonics being generated, and the less symmetrical the device fires, the greater the level of harmonics produced. It is generally undesirable to have high levels of harmonics in a power system.

To help in overcoming this problem, a diac is often placed in series with the gate. This device helps make the switching more even for both halves of the cycle. This results from the fact that the diac switching characteristic is far more even than that of the triac.

Since the diac prevents any gate current flowing until the trigger voltage has reached a certain voltage in either direction, this makes the firing point of the triac more even in both directions. In view of their usefulness, diacs may often be built into the gate terminal of a triac.

TRIAC

DEFINITION

Triacs are widely used in AC power control applications. They are able to switch high voltages and high levels of current, and over both parts of an AC waveform.

This makes triac circuits ideal for use in a variety of applications where power switching is needed. One particular use of triac circuits is in light dimmers for domestic lighting, and they are also used in many other power control situations including motor control.

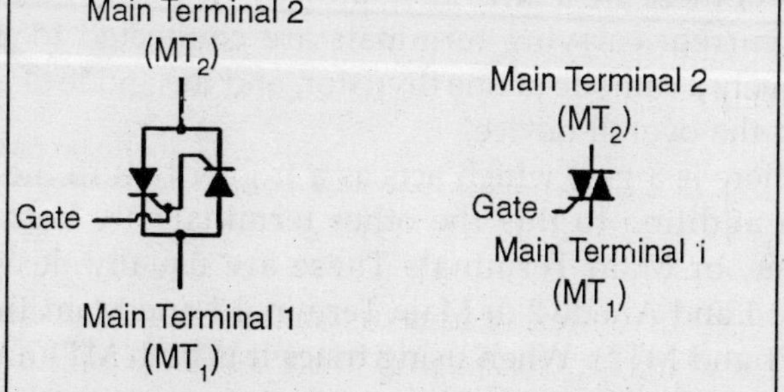

TRIAC Equivalent Circuit TRIAC Schematic Symbol

SCRs are unidirectional (one-way) current devices, making them useful for controlling DC only. If two SCRs are joined in back-to-back parallel fashion just like two Shockley diodes were joined together to form a DIAC, we have a new device known as the *TRIAC*: (Figure above)

The triac is a development of the thyristor. While the thyristor can only control current over one half of the cycle, the triac controls it over two halves of an AC waveform.

As such the triac can be considered as a pair of parallel but opposite thyristors with the two gates connected together and the anode of one device connected to the cathode of the other; etc..

He basic triac symbol used on circuit diagram indicates its bi-directional properties. The triac symbol can be seen to be a couple of thyristor symbols in opposite senses merged together.

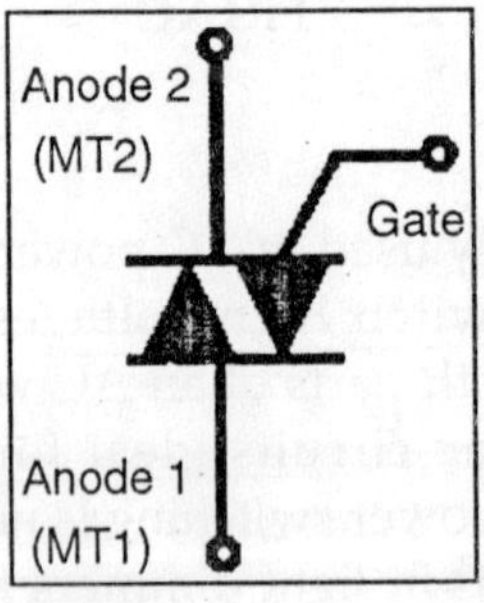

Fig. Triac Symbol for use in Circuit Diagrams

Like a thyristor, a triac has three terminals. However the names of these are a little more difficult to assign, because the main current carrying terminals are connected to what is effectively a cathode of one thyristor, and the anode of another within the overall device.

There is a gate which acts as a trigger to turn the device on. In addition to this the other terminals are both called Anodes, or Main Terminals These are usually designated Anode 1 and Anode 2 or Main Terminal 1 and Main Terminal 2 (MT1 and MT2). When using triacs it is both MT1 and MT2 have very similar properties.

Because individual SCRs are more flexible to use in advanced control systems, these are more commonly seen in circuits like motor drives; TRIACs are usually seen in simple, low-power applications like household dimmer switches. A simple lamp dimmer circuit is shown in Figure below, complete with the phase-shifting resistor-capacitor network necessary for after-peak firing.

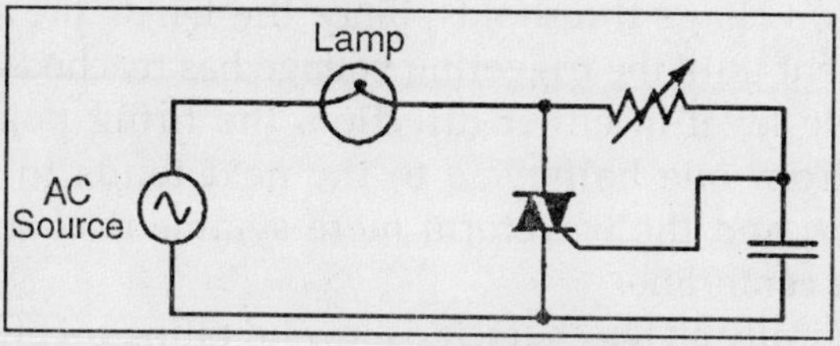

FEATURES

TRIAC Phase-control of Power

TRIACs are notorious for not firing *symmetrically*. This means these usually won't trigger at the exact same gate voltage level for one polarity as for the other. Generally speaking, this is undesirable, because unsymmetrical firing results in a current waveform with a greater variety of harmonic frequencies. Waveforms that are symmetrical above and below their average centerlines are comprised of only odd-numbered harmonics. Unsymmetrical waveforms, on the other hand, contain even-numbered harmonics (which may or may not be accompanied by odd-numbered harmonics as well).

In the interest of reducing total harmonic content in power systems, the fewer and less diverse the harmonics, the better — one more reason individual SCRs are favored over TRIACs for complex, high-power control circuits.

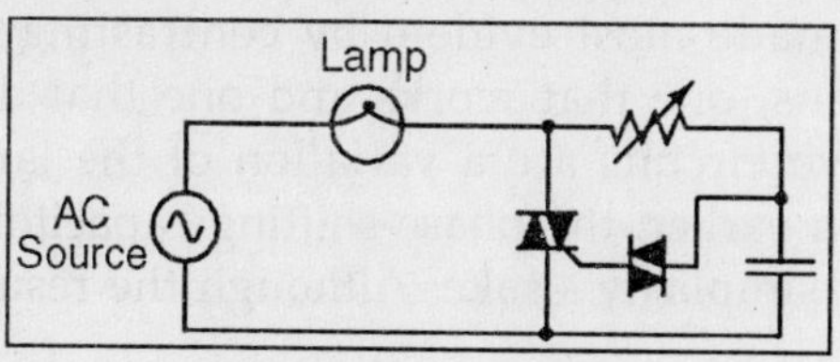

One way to make the TRIAC's current waveform more symmetrical is to use a device external to the TRIAC to time the triggering pulse. A DIAC placed in series with the gate does a fair job of this: (Figure above)

DIAC Improves Symmetry of Control

DIAC breakover voltages tend to be much more symmetrical (the same in one polarity as the other) than TRIAC triggering voltage thresholds. Since the DIAC prevents any gate current until the triggering voltage has reached a certain, repeatable level in either direction, the firing point of the TRIAC from one half-cycle to the next tends to be more consistent, and the waveform more symmetrical above and below its centerline.

Practically all the characteristics and ratings of SCRs apply equally to TRIACs, except that TRIACs of course are bidirectional (can handle current in both directions). Not much more needs to be said about this device except for an important caveat concerning its terminal designations.

From the equivalent circuit diagram shown earlier, one might think that main terminals 1 and 2 were interchangeable. These are not! Although it is helpful to imagine the TRIAC as being composed of two SCRs joined together, it in fact is constructed from a single piece of semiconducting material, appropriately doped and layered.

The actual operating characteristics may differ slightly from that of the equivalent model.

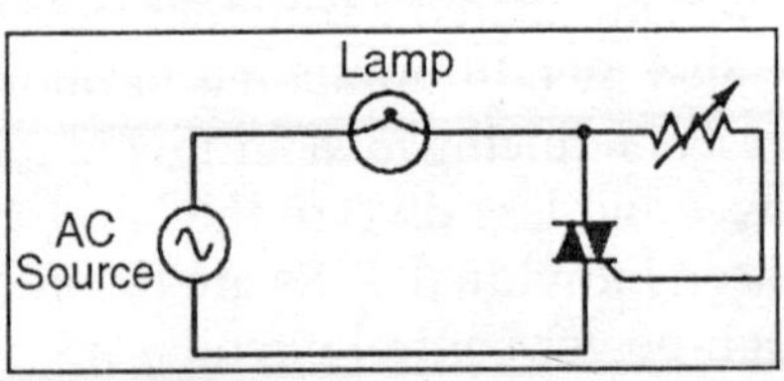

This is made most evident by contrasting two simple circuit designs, one that works and one that doesn't. The following two circuits are a variation of the lamp dimmer circuit shown earlier, the phase-shifting capacitor and DIAC removed for simplicity's sake. Although the resulting circuit lacks the fine control ability of the more complex version (with capacitor and DIAC), it *does* function: (Figure above)

This circuit with the gate to MT_2 *does function.* Suppose we were to swap the two main terminals of the TRIAC around. According to the equivalent circuit diagram shown earlier in this section, the swap should make no difference. The circuit ought to work: (Figure below)

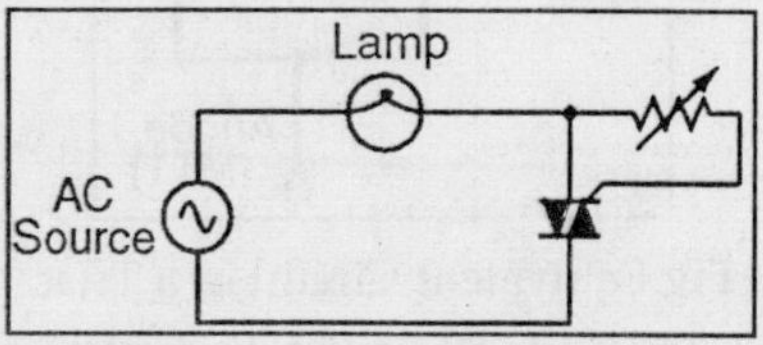

With the gate swapped to MT_1*, this circuit does not function.* However, if this circuit is built, it will be found that it does not work! The load will receive no power, the TRIAC refusing to fire at all, no matter how low or high a resistance value the control resistor is set to.

The key to successfully triggering a TRIAC is to make sure the gate receives its triggering current from the *main terminal 2* side of the circuit (the main terminal·on the opposite side of the TRIAC symbol from the gate terminal). Identification of theM T_1 and MT_2 terminals must be done via the TRIAC's part number with reference to a data sheet or book.

WORKING

Before looking at how a triac works, it helps to have an understanding of haow a thyristor works. In this way the basic concepts can be grasped for the simpler device and then applied to a triac which is more complicated. The operation of the thyristor is covered in the article in this section and accessible through the "Related Articles" box on the left of the page and below the main menu.

For the operation of the triac, it can be imagined from the circuit symbol that the triac consists of two thyristors in parallel but around different ways. The operation of the triac can be looked on in this fashion, although the actual operation at the semiconductor level is rather more complicated.

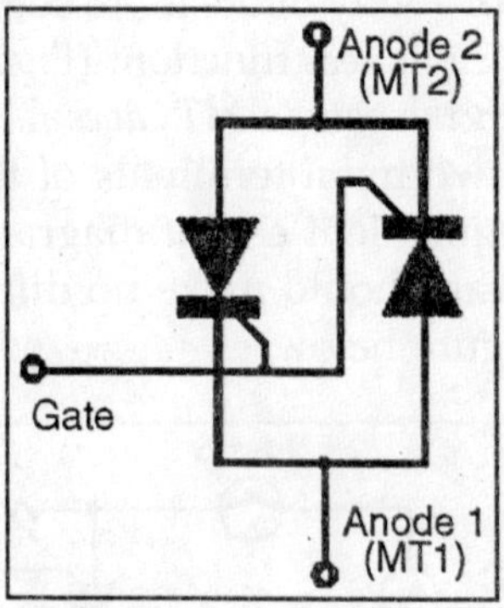

Fig.Equivalent Circuit of a Triac

When the voltage on the MT1 is positive with regard to MT2 and a positive gate voltage is applied, one of the thyristors conducts. When the voltage is reversed and a negative voltage is applied to the gate, the other thyristor conducts. This is provided that there is sufficient voltage across the device to enable a minimum holding current to flow.

APPLICATIONS

Although these devices operate very well, to get the best performance out of them it is necessary to understand a few hints on tips on using triacs. It is found that because of their internal construction and the slight differences between the two halves, triacs do not fire symmetrically. This results in harmonics being generated: the less symmetrical the triac fires, the greater the level of harmonics that are produced.

It is not normally desirable to have high levels of harmonics in a power system and as a result triacs are not favoured for high power systems. Instead for these systems two thyristors may be used as it is easier to control their firing.

To help in overcoming the problem non-symmetrical firing ad the resulting harmonics, a device known as a diac (diode AC switch) is often placed in series with the gate of the triac. The inclusion of this device helps make the switching more even for both halves of the cycle. This results from the fact that the diac switching characteristic is far more even than that of the triac.

Chapter 6

Operational Amplifier

INVERTING AND NON-INVERTING CONFIGURATIONS

INTRODUCTION

The operational amplifier is arguably the most useful single device in analog electronic circuitry. With only a handful of external components, it can be made to perform a wide variety of analog signal processing tasks. It is also quite affordable, most general-purpose amplifiers selling for under a dollar apiece. Modern designs have been engineered with durability in mind as well: several "op-amps" are manufactured that can sustain direct short-circuits on their outputs without damage.

One key to the usefulness of these little circuits is in the engineering principle of feedback, particularly *negative* feedback, which constitutes the foundation of almost all automatic control processes. The principles presented here in operational amplifier circuits, therefore, extend well beyond the immediate scope of electronics. It is well worth the electronics student's time to learn these principles and learn them well.

INVERTING AMPLIFIER

The inverting amplifier is a particularly useful form of circuit to be able to use. It can take many forms, but when talking of an inverting amplifier, its realisation is normally associated with the use of operational amplifiers.

Operational amplifiers or op-amps are a key element analogue circuit designers library. Operational amplifiers are very easy to use and they offer a very high level of performance. In addition to this, operational amplifier integrated circuits are very cheap and easy to obtain.

Apart from their use in inverting amplifier configurations, operational amplifiers for the basis of a very wide variety of circuits from astables and multivibrators to filters, regulators, differentiators, and very many more types of circuit.

Operational amplifiers have been available in the form of integrated circuits for very many years. They are available in a wide variety of packages. Not only are they available in dual in line packages with one, two, or more amplifiers in a single package, but they are also available in surface mount varieties as well.

Inverting amplifier circuit: Here the output is an inverted version (upside down) of the signal that enters the circuit, i.e. when the input voltage goes more positive, the output goes more negative.

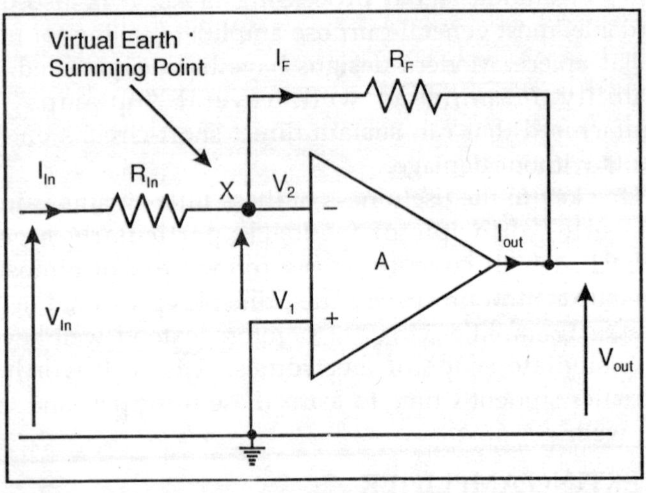

In this Inverting Amplifier circuit the operational amplifier is connected with feedback to produce a closed loop operation. There are two very important rules to remember about inverting amplifiers is that, "no current flows into the input terminal" and that "V1 equals V2". This is because the junction of the input and feedback signal (×) is at the same potential as the positive (+) input which is at zero volts or ground then, the junction is a "Virtual Earth". Because of this virtual earth node the input resistance of the amplifier is equal to the value of the input resistor, Rin and the closed loop gain of the inverting amplifier can be set by the ratio of the two external resistors.

We said above that there are two very important rules to remember about Inverting Amplifiers or any operational amplifier for that matter and they are.

- No Current Flows into the Input Terminals
- The Differential Input Voltage is Zero as V1 = V2 = 0 (Virtual Earth)

Then by using these two rules we can find the equation for calculating the gain of an inverting amplifier, using first principles.

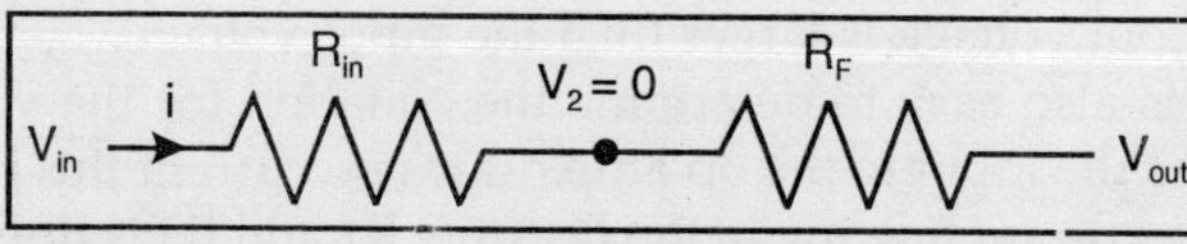

Fig. Current (i) Flows Through the Resistor Network as Shown

$$i = \frac{Vin - Vout}{Rin + Rf}$$

$$\text{There fore,i} = \frac{Vin - V2}{Rin} = \frac{V2 - Vout}{Rf}$$

$$i = \frac{Vin}{Rin} - \frac{V2}{Rin} = \frac{V2}{Rf} - \frac{Vout}{Rf}$$

$$so, \frac{Vin}{Rin} = V2\left[\frac{1}{Rin} + \frac{1}{Rf}\right] - \frac{Vout}{Rf}$$

$$\text{and as, i} = \frac{Vin - 0}{Rin} = \frac{0 - Vout}{Rf} \frac{Rf}{Rin} = \frac{0 - Vout}{Vin - 0}$$

$$\text{The Closed Loop Gain is given as,} \frac{Vout}{Vin} = -\frac{Rf}{Rin}$$

Then, the Closed-Loop Voltage Gain of an Inverting Amplifier is given as.

$$\text{Gain}=\frac{\text{Vout}}{\text{Vin}}=-\frac{Rf}{Rin}$$

and this can be transposed to give:

$$\text{Vout}=-\frac{Rf}{Rin}\times Vin$$

The negative sign in the equation indicates an inversion of the output signal with respect to the input as it is 180° out of phase. This is due to the feedback being negative in value.

Inverting Amplifier Gain

One of the main features of the inverting amplifier circuit is the overall gain that it produces. This is quite easy to calculate. The voltage gain is actually the output voltage (Vout) divided by the input voltage (Vin), i.e. it is the number of times the output voltage is larger than the input voltage.

It is also easy to determine the equation for the voltage gain. As the input to the op-amp draws no current this means that the current flowing in the resistors R1 and R2 is the same. Using ohms law Vout/R2 = –Vin/R1. Hence the voltage gain of the circuit Av can be taken as:

Av = –R2/R1

As an example, an amplifier requiring a gain of ten could be built by making R2 47 k ohms and R1 4.7 k ohms.

Inverting Amplifier Input Impedance

It is often necessary to know the input impedance of a circuit, and in this case of the inverting amplifier. A circuit with a low input impedance may load the output of the previous circuit and may give rise to effects such as changing the frequency response if the coupling capacitors are not large.

It is very simple to determine the input impedance of an inverting operational amplifier circuit. It is simply the value of the input resistor R1. This is because the inverting input is at earth potential (i.e. a virtual earth) and this means that the resistor is connected between the input and earth.

Example 1: Find the closed loop gain of the following inverting amplifier circuit.

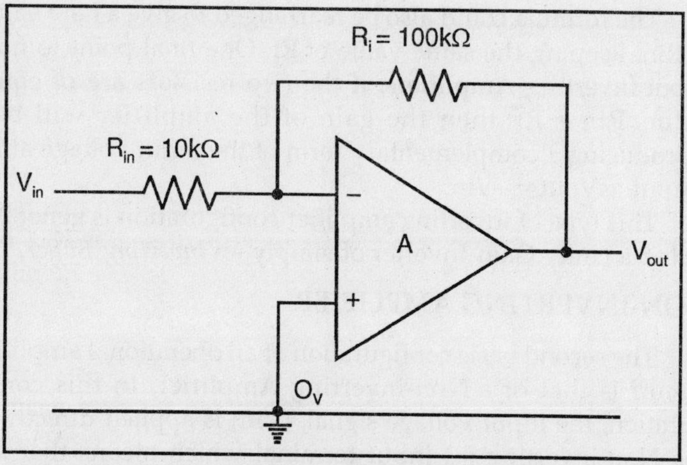

Using the previously found formula for the gain of the circuit

$$\text{Gain}=\frac{\text{Vout}}{\text{Vin}}=-\frac{Rf}{Rin}$$

we can now substitute the values of the resistors in the circuit as follows,

Rin = 10kΩ and Rf = 100kΩ.

and the gain of the circuit is calculated as,

–Rf/Rin = 100k/10k = 10.

Therefore, the closed loop gain of the inverting amplifier circuit above is given 10 or 20dB.

Example 2: The gain of the original circuit is to be increased to 40, find the new values of the resistors required.

Assume that the input resistor is to remain at the same value of 10KΩ, then by re-arranging the closed loop voltage gain formula we can find the new value required for the feedback resistor Rf.

Gain = –Rf/Rin

therefore, Rf = Gain × Rin

Rf = 40 × 10,000

Rf = 400,000 or 400KΩ

The new values of resistors required for the circuit to have a gain of 40 would be, Rin = 10KΩ and Rf = 400KΩ.

The formula could also be rearranged to give a new value of Rin, keeping the same value of Rf. One final point to note about Inverting Amplifiers, if the two resistors are of equal value, Rin = Rf then the gain of the amplifier will be -1 producing a complementary form of the input voltage at its output asVout = –Vin.

This type of inverting amplifier configuration is generally called a Unity Gain Inverter of simply an *Inverting Buffer*.

NON-INVERTING AMPLIFIER

The second basic configuration of an operational amplifier circuit is that of a Non-inverting Amplifier. In this configuration, the input voltage signal, (Vin) is applied directly to the Non-inverting (+) input terminal which means that the output gain of the amplifier becomes "Positive" in value in contrast to the "Inverting Amplifier" circuit we saw in the last tutorial and whose output gain is negative in value.

Feedback control of the non-inverting amplifier is achieved by applying a small part of the output voltage signal back to the inverting (–) input terminal via a Rf – R2 voltage divider network, again producing negative feedback.

This produces a Non-inverting Amplifier circuit with very good stability, a very high input impedance, Rinapproaching infinity (as no current flows into the positive input terminal) and a low output impedance,Rout as shown below.

Circuit

In the previous Inverting Amplifier tutorial, we said that "no current flows into the input" of the amplifier and that "V1 equals V2". This was because the junction of the input and feedback signal (V1) are at the same potential in other words the junction is a "Virtual Earth" summing point.

Because of this virtual earth node the resistors, Rf and R2 form a simple voltage divider network across the amplifier and the voltage gain of the circuit is determined by the ratios of R2 and Rf as shown above.

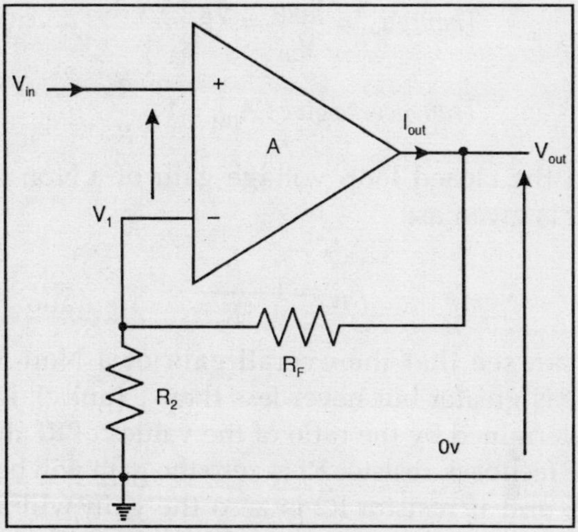

Equivalent Voltage Divider Network

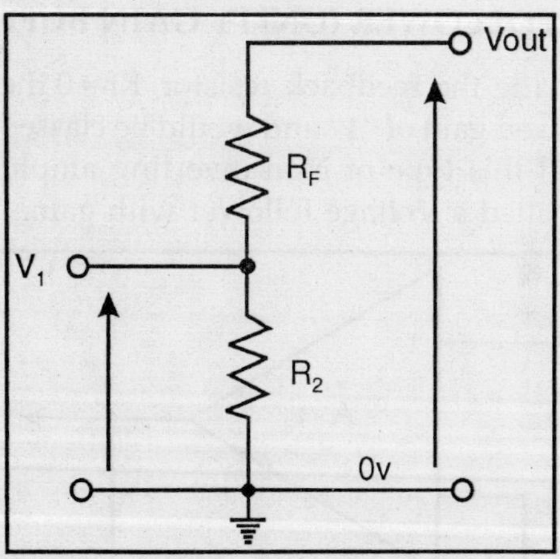

Then using the formula to calculate the output voltage of a potential divider network, we can calculate the output Voltage Gain of the Non-inverting Amplifier as:

$$V_1 = \frac{R_2}{R_2 + R_F} \times V_{OUT}$$

$$Voltage\,Gain, A_{(V)}\ is\ equal\ to : \frac{V_{OUT}}{V_{IN}}$$

$$Then, A_{(V)} = \frac{V_{OUT}}{V_{IN}} = \frac{R_R + R_F}{R_2}$$

$$Transpose\ to\ give : A_{(V)} = 1 + \frac{R_F}{R_2}$$

Then the closed loop voltage gain of a Non-inverting Amplifier is given as:

$$A_{(V)} = 1 + \frac{R_F}{R_2}$$

We can see that the overall gain of a Non-Inverting Amplifier is greater but never less than 1 (unity), is positive and is determined by the ratio of the values of Rf and R2.

If the feedback resistor Rf is zero the gain will be equal to 1 (unity), and if resistor R2 is zero the gain will approach infinity, but in practice it will be limited to the operational amplifiers open-loop differential gain, (A_o).

VOLTAGE FOLLOWER (UNITY GAIN BUFFER)

If we made the feedback resistor, Rf = 0 then the circuit will have a fixed gain of "1" and would be classed as a Voltage Follower and this type of Non-inverting amplifier circuit is sometimes called a Voltage follower with gain.

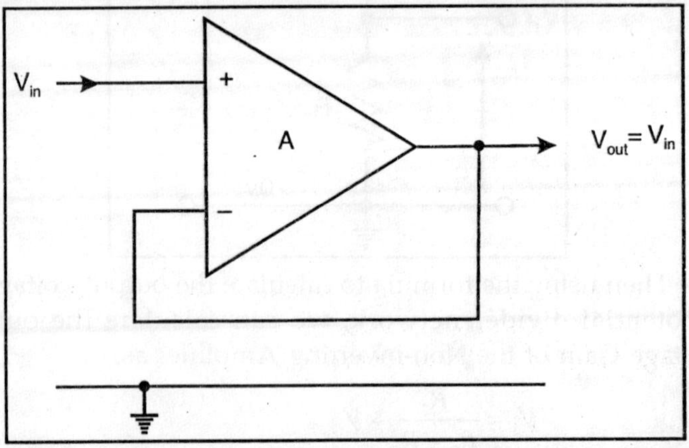

As the input signal is connected directly to the non-inverting input of the amplifier the output signal is not inverted resulting in the output voltage being equal to the input voltage, Vout = Vin.

This then makes the Voltage Follower circuit ideal as a Unity Gain Buffer circuit because of its isolation properties as impedance or circuit isolation is more important than amplification.

The input impedance of the voltage follower circuit is very high, typically above 1MΩ as it is equal to that of the operational amplifiers input, since an ideal op-amp condition is assumed.

In this circuit, Rin has increased to infinity and Rf reduced to zero, the feedback is 100% and Vout is exactly equal to Vin giving it a fixed gain of 1 or unity. As the input voltage Vin is applied to the non-inverting input the gain of the amplifier is given as:

$$A_{out} = A(V_{in} - V_{out})$$

$$as V_{in} = V+ and V_{out} = V-$$

$$there fore, Gain = \frac{V_{out}}{V_{in}} = 1$$

Since no current flows into the non-inverting input terminal the input impedance is infinite and also no current flows through the feedback loop so any value of resistance may be placed in the feedback loop without affecting the characteristics of the circuit as no voltage is dissipated across it, zero current, zero voltage, zero power drop.

However in most real unity gain buffer circuits a low value (typically 1kΩ) resistor is required to reduce any offset input leakage currents, and also if the operational amplifier is of a current feedback type.

The voltage follower or unity gain buffer is a special and very useful type of Non-inverting amplifiercircuit that is commonly used in electronics to isolated circuits from each other especially in High-order state variable or Sallen-Key type active filters to separate one filter stage from the other. Typical digital buffer IC's available are the 74LS125 Quad 3-state buffer or the more common 74LS244 Octal buffer.

One final thought, the output voltage gain of the voltage follower circuit with closed loop gain is Unity, the voltage gain of an ideal operational amplifier with open loop gain (no feedback) is Infinite.

Then by carefully selecting the feedback components we can control the amount of gain produced by an Operational Amplifier anywhere from 1 to infinity.

REVISION QUESTIONS

Question 1: Write the transfer function (input/output equation) for an operational amplifier with an open-loop voltage gain of 100,000, and the inverting input connected directly to its output terminal. In other words, write an equation describing the output voltage of this op-amp (V_{out}) for any given input voltage at the noninverting input ($V_{in(+)}$):

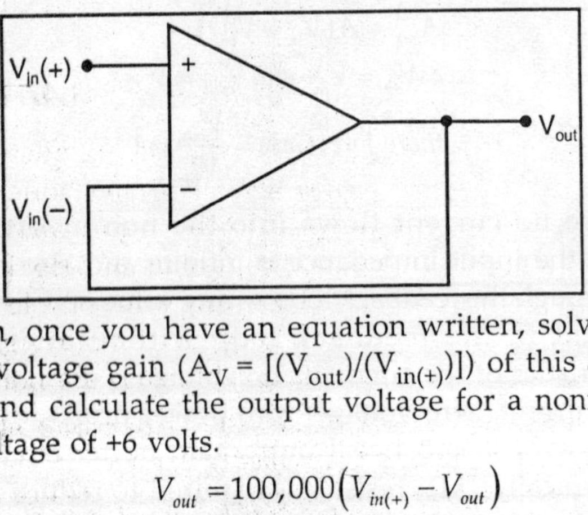

Then, once you have an equation written, solve for the over-all voltage gain ($A_V = [(V_{out})/(V_{in(+)})]$) of this amplifier circuit, and calculate the output voltage for a noninverting input voltage of +6 volts.

$$V_{out} = 100{,}000\left(V_{in(+)} - V_{out}\right)$$

$$A_V = \frac{100{,}000}{100{,}001} = 0.9999$$

For an input voltage of +6 volts, the output voltage will be +5.99994 volts.

The significant point of this question is that students see the over-all voltage gain of the opamp radically attenuated from 100,000 to approximately 1. What is not so evident is just how *stable*this new voltage gain is, which is one of the purposes for employing negative feedback.

Question 2: Write the transfer function (input/output equation) for an operational amplifier with an open-loop voltage gain of 100,000, and the inverting input connected to a voltage divider on its output terminal (so the inverting input receives exactly one-half the output voltage). In other words, write an equation describing the output voltage of this op-amp (V_{out}) for any given input voltage at the noninverting input ($V_{in(+)}$):

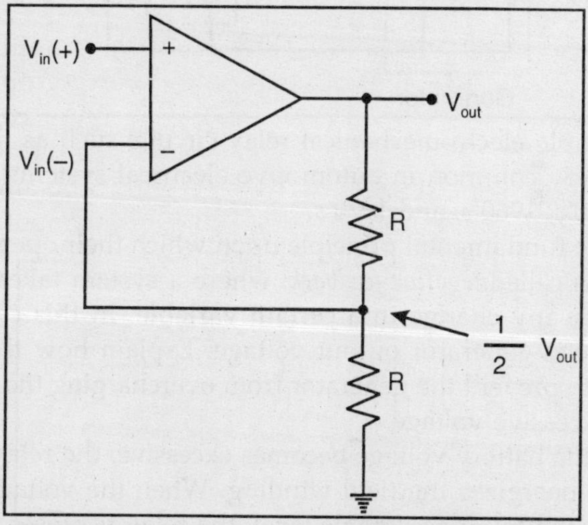

Then, once you have an equation written, solve for the output voltage if the noninverting input voltage is –2.4 volts.

$$V_{out} = 100{,}000(V_{in(+)} - {}^{1}/_{2}V_{out})$$

I've left it up to you to perform the algebraic simplification here!

For an input voltage of –2.4 volts, the output voltage will be –4.7999 volts.

Follow-up question: what do you notice about the output voltage in this circuit? What value is it very close to being, in relation to the input voltage? Does this pattern hold true for other input voltages as well?

Question 3: Generators used in battery-charging systems must be *regulated* so as to not overcharge the battery(ies) they are connected to. Here is a crude, relay-based voltage regulator for a DC generator:

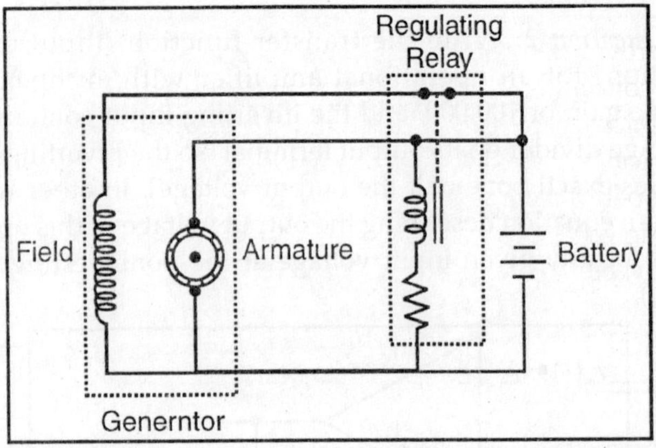

Simple electromechanical relay circuits such as this one were very common in automotive electrical systems during the 1950's, 1960's, and 1970's.

The fundamental principle upon which their operation is based is called*negative feedback*: where a system takes action to *oppose* any change in a certain variable. In this case, the variable is generator output voltage. Explain how the relay works to prevent the generator from overcharging the battery with excessive voltage.

If the battery voltage becomes excessive, the relay opens and de-energizes the field winding. When the voltages sags back down to an acceptable level, the relay re-closes and re-energizes the field winding so that the generator can begin generating voltage again.

Challenge question: what would we have to change in this circuit to alter the generator's voltage regulation set-point (the "target" voltage at which the generator's output is supposed to be regulated)?

The circuit drawn here is very similar to real generator regulator circuits used in American automobiles before the advent of inexpensive, reliable semiconductor circuits. I show it here not just for historical background, but also to demonstrate how relatively crude circuits are still able to perform certain tasks reasonably well.

"Negative feedback" is one of the fundamental principles of electronics and electrical engineering. A simple system like this provides a good way to gently introduce students to this vital concept.

Question 4: A mechanic has an idea for upgrading the electrical system in an automobile originally designed for 6 volt operation.

He wants to upgrade the 6 volt headlights, starter motor, battery, etc, to 12 volts, but wishes to retain the original 6-volt generator and regulator. Shown here is the original 6-volt electrical system:

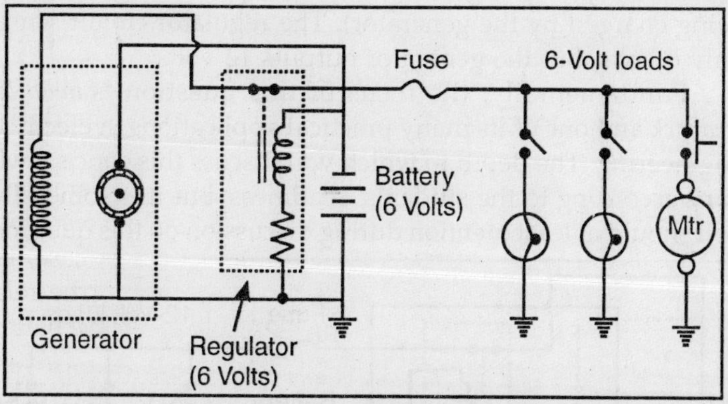

The mechanic's plan is to replace all the 6-volt loads with 12-volt loads, and use two 6-volt batteries connected in series, with the original (6-volt) regulator sensing voltage across only one of those batteries:

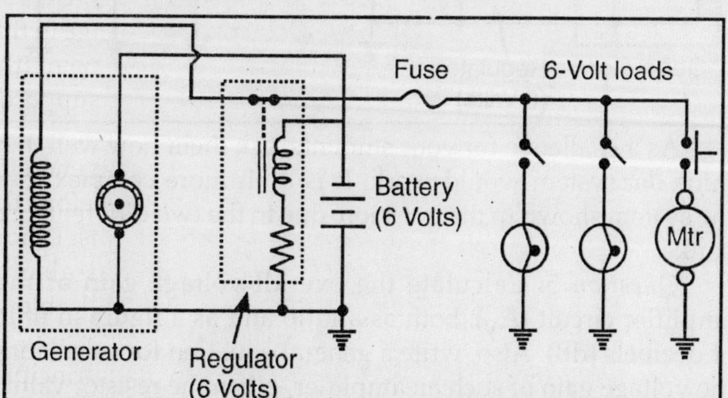

Explain how this system is supposed to work. Do you

think the mechanic's plan is practical, or are there any problems with it?

So long as the generator is capable of outputting 12 volts, this system will work!

Challenge question: identify factors that may prevent the generator from outputting enough voltage with the regulator connected as shown in the last diagram.

In this question, we see a foreshadowing of op-amp theory, with the regulator's negative feedback applied to what is essentially a voltage divider (two equal-voltage batteries being charged by the generator). The regulator circuit senses only 6 volts, but the generator outputs 12 volts.

Fundamentally, the focus of this question is *negative feedback* and one of its many practical applications in electrical engineering. The depth to which you discuss this concept will vary according to the students' readiness, but it is something you should at least mention during discussion on this question.

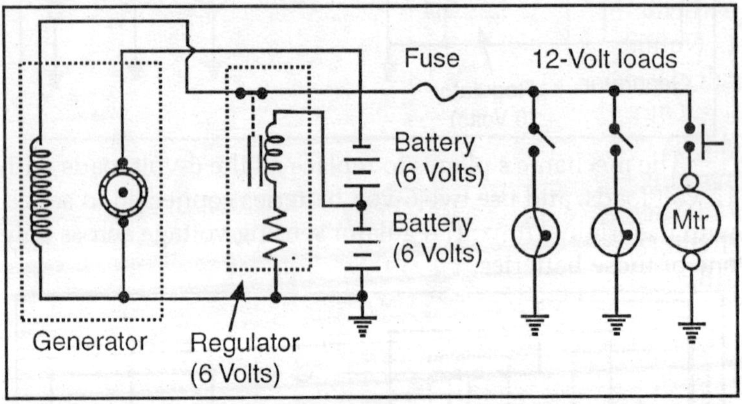

As a challenge for your students, ask them how well they think *this* system would work. It is a bit more complex than the system shown in the question, due to the two different load banks.

Question 5: Calculate the overall voltage gain of this amplifier circuit (A_V), both as a ratio and as a figure in units of decibels (dB). Also, write a general equation for calculating the voltage gain of such an amplifier, given the resistor values of R_1 and R_2:

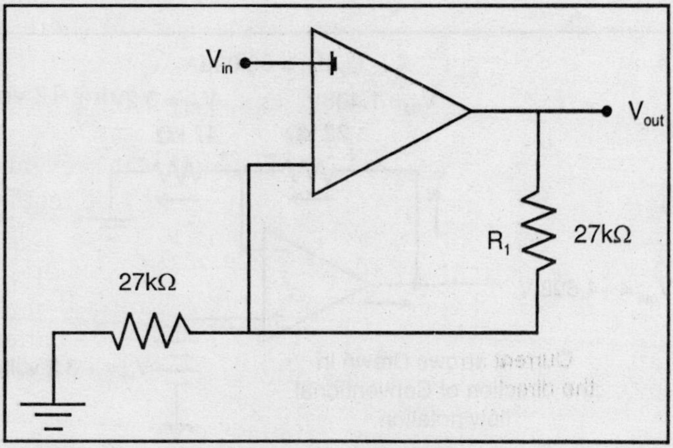

$A_V = 2 = 6.02$ dB

$$A_V = \frac{R_1}{R_2} + 1 \quad \text{(expressed as a ratio, not dB)}$$

Follow-up question: explain how you could modify this particular circuit to have a voltage gain (ratio) of 3 instead of 2. Nothing special here - just some practice with voltage gain calculations.

Question 6: Calculate all voltage drops and currents in this circuit, complete with arrows for current direction and polarity markings for voltage polarity.

Then, calculate the overall voltage gain of this amplifier circuit (A_V), both as a ratio and as a figure in units of decibels (dB):

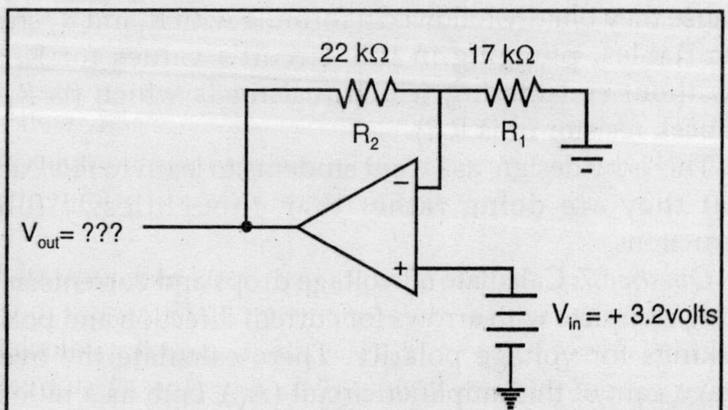

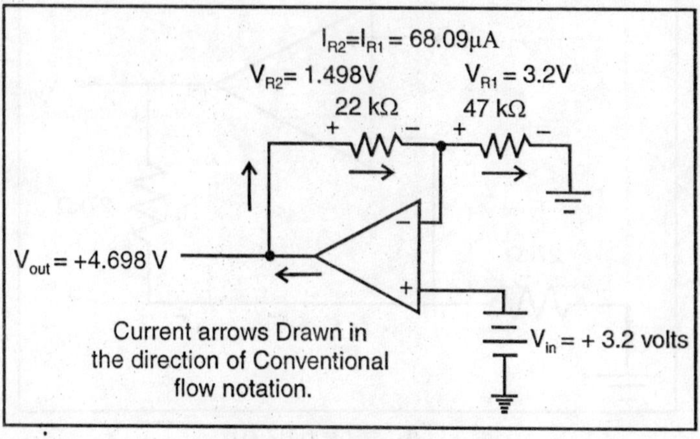

$$A_V = 1.468 = 3.335 \text{ dB}$$

Operational amplifier circuits provide a great opportunity to review basic concepts of DC circuits: voltage drops, polarity, current directions, Ohm's Law, Kirchhoff's Voltage Law, Kirchhoff's Current Law, etc.

This circuit is no exception. Emphasize the fact that a great many opamp circuits may be comprehensively analyzed merely with knowledge of these fundamental principles and the characteristics of an ideal opamp (zero input current, infinite open-loop gain, unlimited output voltage swing, zero voltage between input terminals when negative feedback is in effect).

Some students may arrive at the wrong gain figure because they blindly followed a formula with $R_1$and R_2 shown as variables, plugging in this circuit's values for R_1 and R_2 without considering which resistor is which (is R_1 the feedback resistor or is R_2?).

This is by design, as I want students to learn to *think* about what they are doing rather than thoughtlessly follow instructions.

Question 7: Calculate all voltage drops and currents in this circuit, complete with arrows for current direction and polarity markings for voltage polarity. Then, calculate the overall voltage gain of this amplifier circuit (A_V), both as a ratio and as a figure in units of decibels (dB):

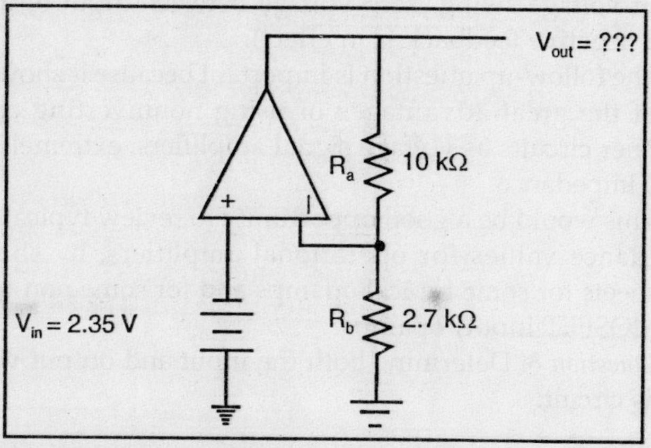

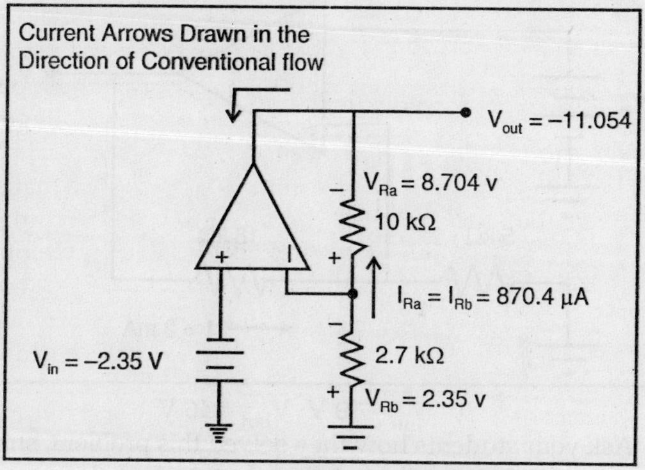

$A_V = 4.704 = 13.449$ dB

Follow-up question: how much input impedance does the -2.35 volt source ßee" as it drives this amplifier circuit?

Notes: Operational amplifier circuits provide a great opportunity to review basic concepts of DC circuits: voltage drops, polarity, current directions, Ohm's Law, Kirchhoff's Voltage Law, Kirchl.off's Current Law, etc. This circuit is no exception. Emphasize the fact that a great many opamp circuits may be comprehensively analyzed merely with knowledge of these fundamental principles and the characteristics of an ideal opamp (zero input current, infinite open-loop gain, unlimited output voltage swing, zero voltage between input terminals when negative feedback is in effect).

The follow-up question is important because it showcases one of the great advantages of using noninverting opamp amplifier circuits as voltage signal amplifiers: extremely high input impedance.

This would be a good opportunity to review typical input impedance values for operational amplifiers, by showing datasheets for some typical opamps and for some non-typical (i.e. MOSFET input) opamps.

Question 8: Determine both the input and output voltage in this circuit:

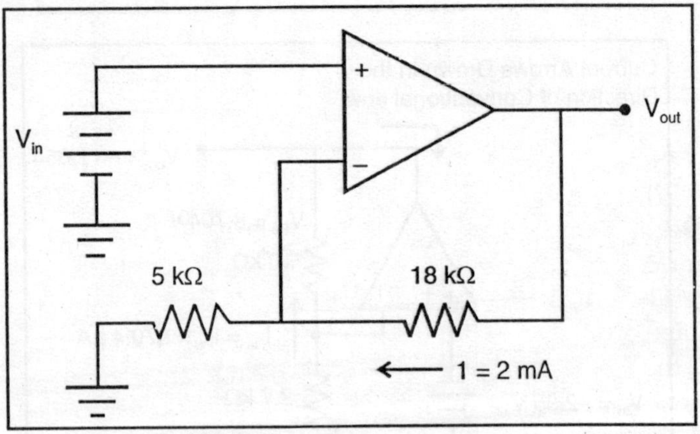

$V_{in} = 10$ V $V_{out} = 46$ V

Ask your students how they solved this problem, sharing techniques and strategies to help other students know where to begin and where to proceed from there.

Question 9: What would have to be altered in this circuit to increase its overall voltage gain?

The voltage divider would have to altered so as to send a smaller proportion of the output voltage to the inverting input.

Ask your students to explain how they would modify the voltage divider in this circuit to achieve the goal of a smaller voltage division ratio. This should be trivial, but it is always good to review basic principles of electricity even when "deep" into a more advanced topic.

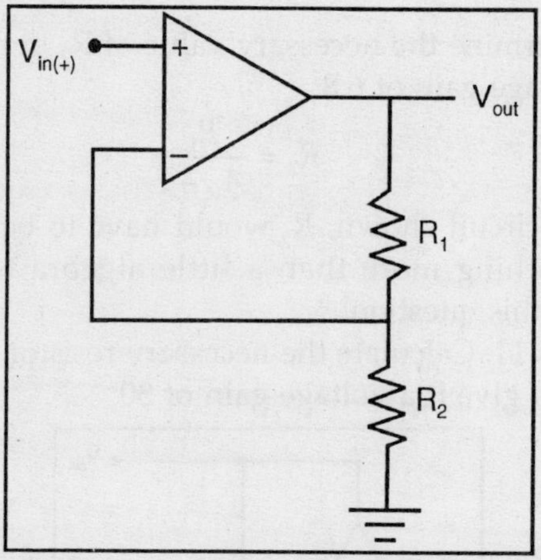

Question 10: The equation for voltage gain (A_V) in a typical no inverting, single-ended opamp circuit is as follows:

$$A_V = \frac{R_1}{R_2} + 1$$

Where,

R_1 is the feedback resistor (connecting the output to the inverting input) R_2 is the other resistor (connecting the inverting input to ground) Suppose we wished to change the voltage gain in the following circuit from 5 to 6.8, but only had the freedom to alter the resistance of R_2:

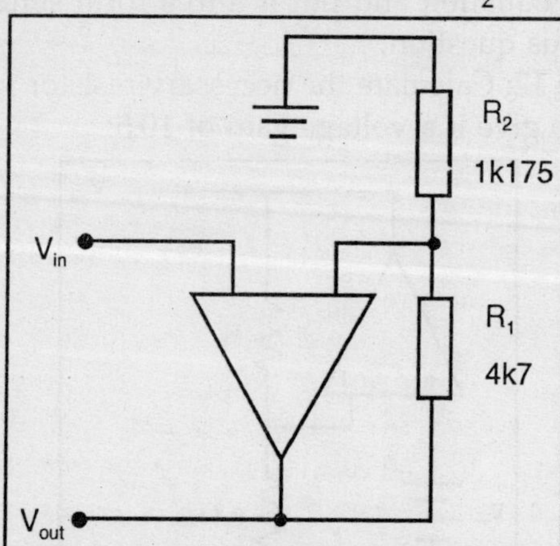

Algebraically manipulate the gain equation to solve for

R_2, then determine the necessary value of R_2 in this circuit to give it a voltage gain of 6.8.

$$R_2 = \frac{R_1}{A_V - 1}$$

For the circuit shown, R_2 would have to be set equal to 810.3 Ω. Nothing more than a little algebra to obtain the answers for this question!

Question 11: Calculate the necessary resistor value (R_1) in this circuit to give it a voltage gain of 30:

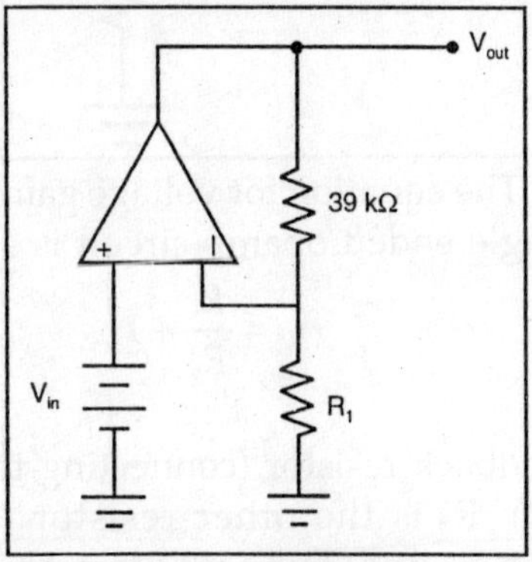

R_1 = 1.345 kΩ

Algebraic manipulation is necessary to take the standard voltage gain equation and put it into a form suitable for use answering this question.

Question 12: Calculate the necessary resistor value (R_1) in this circuit to give it a voltage gain of 10.5:

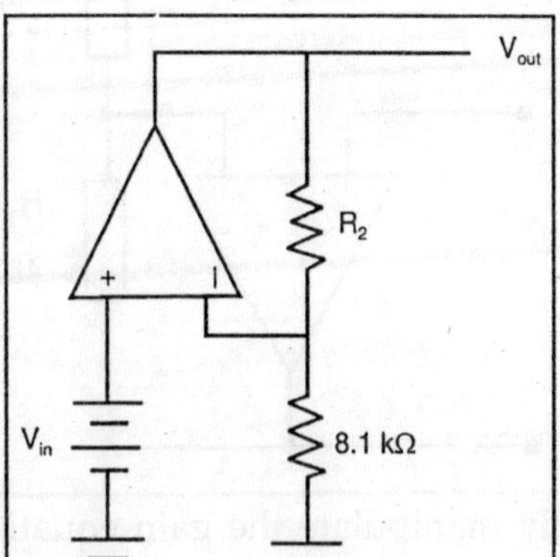

$$R_1 = 76.95\ k\Omega$$

Algebraic manipulation is necessary to take the standard voltage gain equation and put it into a form suitable for use answering this question.

Question 13: Predict how the operation of this operational amplifier circuit will be affected as a result of the following faults. Consider each fault independently (i.e. one at a time, no multiple faults):

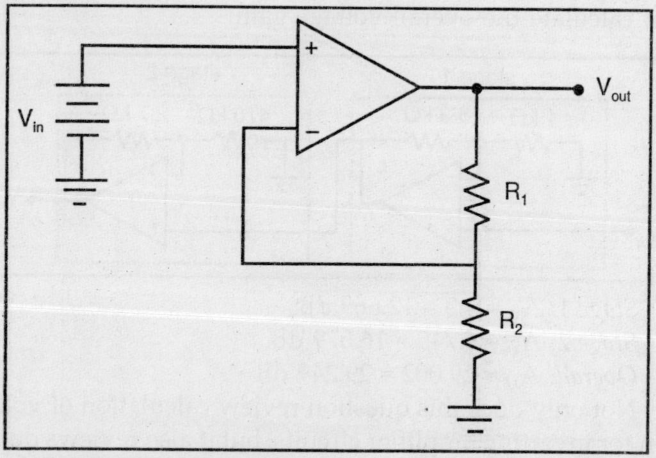

- Resistor R_1 fails open:
- Solder bridge (short) across resistor R_1:
- Resistor R_2 fails open:
- Solder bridge (short) across resistor R_2:

Broken wire between R_1/R_2 junction and inverting opamp input:

- Resistor R_1 fails open: *Output saturates positive.*
- Solder bridge (short) across resistor R_1: $V_{out} = V_{in}$.
- Resistor R_2 fails open: $V_{out} = V_{in}$.

Solder bridge (short) across resistor R_2: *Output saturates positive.*

Broken wire between R_1/R_2 junction and inverting opamp input: *Output voltage unpredictable.*

Notes: The purpose of this question is to approach the domain of circuit troubleshooting from a perspective of knowing what the fault is, rather than only knowing what the symptoms are. Although this is not necessarily a realistic perspective, it helps students build the foundational knowledge necessary to diagnose a faulted circuit from empirical data. Questions such as this should be followed (eventually) by other questions asking students to identify likely faults based on measurements.

Question 14: Calculate the voltage gain for each stage of this amplifier circuit (both as a ratio and in units of decibels), then calculate the overall voltage gain:

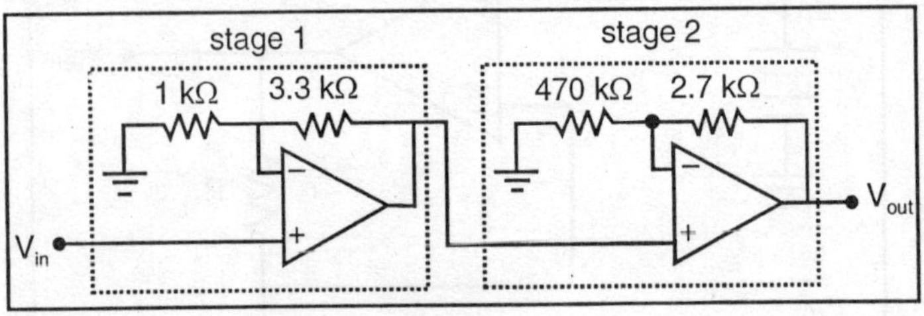

Stage 1: $A_V = 4.3 = 12.669$ dB
Stage 2: $A_V = 6.745 = 16.579$ dB
Overall: $A_V = 29.002 = 29.249$ dB

Not only does this question review calculation of voltage gain for inverting amplifier circuits, but it also reviews decibel calculations (for both single and multi-stage amplifiers). Discuss how the decibel figures for each stage add to equal the total decibel gain, whereas the ratios multiply.

Question 15: Calculate the voltage gain for each stage of this amplifier circuit (both as a ratio and in units of decibels), then calculate the overall voltage gain:

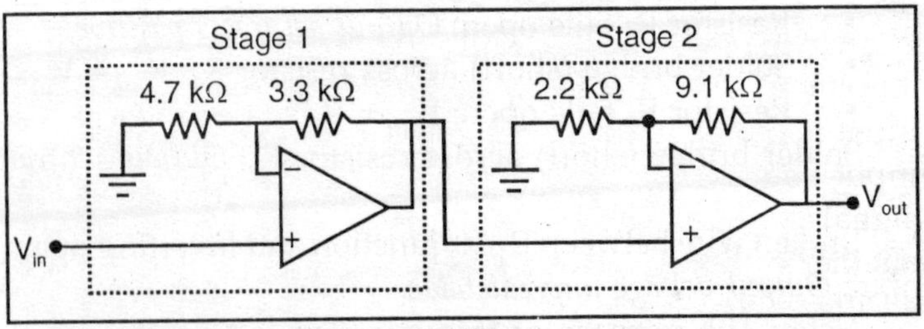

Stage 1: $A_V = 1.702 = 4.62$ dB
Stage 2: $A_V = 5.136 = 14.213$ dB
Overall: $A_V = 8.743 = 18.833$ dB

Not only does this question review calculation of voltage gain for inverting amplifier circuits, but it also reviews decibel calculations (for both single and multi-stage amplifiers). Discuss how the decibel figures for each stage add to equal the total decibel gain, whereas the ratios multiply.

Question 16: How much effect will a change in the op-amp's open-loop voltage gain have on the *overall* voltage gain of a negative-feedback circuit such as this?

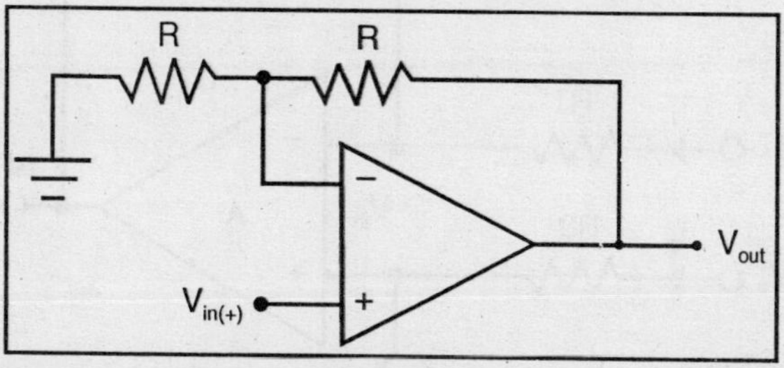

If the open-loop gain of this operational amplifier were to change from 100,000 to 200,000, for example, how big of an effect would it have on the voltage gain as measured from the noninverting input to the output?

The different in overall voltage gain will be trivial.

Follow-up question: what advantage is there in building voltage amplifier circuits in this manner, applying negative feedback to a "core" amplifier with very high intrinsic gain?

DIFFERENCE AMPLIFIER

DEFINITION

Up to now we have used only one input to connect to the amplifier, using either the "Inverting" or the "Non-inverting" input terminal to amplify a single input signal with the other input being connected to ground. But we can also connect signals to both of the inputs at the same time producing another common type of operational amplifier circuit called a Differential Amplifier.

By connecting one voltage signal onto one input terminal and another voltage signal onto the other input terminal the resultant output voltage will be proportional to the

"Difference" between the two input signals of V1 and V2 and this type of circuit can also be used as a Subtractor. Then, this type of Operational Amplifier circuit is commonly known as a Differential Amplifier configuration and is shown below:

CONSTRUCTION

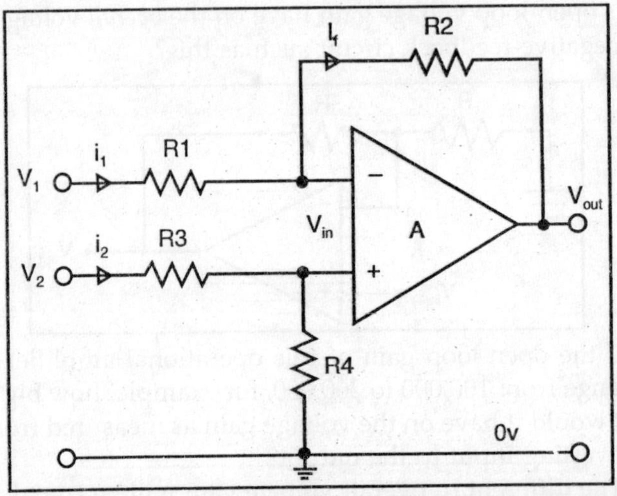

The transfer function for a Differential Amplifier circuit is given as:

$$V_{OUT} = \frac{R_2}{A_2} + V_1 + \left(1 + \frac{R_2}{R_1}\right)\left(\frac{R_4}{R_3 + R_4}\right)V_2$$

When R1 = R3 and R2 = R4 the transfer function formula can be modified to the following:

$$V_{OUT} = \frac{R_2}{R_1}(V_2 - V_1)$$

If all the resistors are all of the same ohmic value the circuit will become a Unity Gain Differential Amplifier and the gain of the amplifier will be 1 or Unity.

The Differential Amplifier circuit is a very useful op-amp circuit and by adding more resistors in parallel with the input resistors R1 and R3, the resultant circuit can be made to either "Add" or "Subtract" the voltages applied to their respective inputs. One of the most common ways of doing this is to connect a "Resistive Bridge" commonly called a *Wheatstone Bridge* to the input of the amplifier.

FEATURES

As Bridge Amplifier

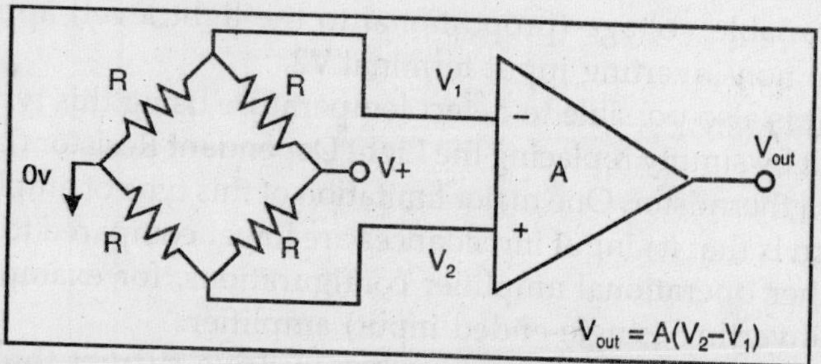

The standard Differential Amplifier circuit now becomes a differential voltage comparator by "Comparing" one input voltage to the other.

For example, by connecting one input to a fixed voltage reference set up on one leg of the resistive bridge network and the other to either a "Thermistor" or a "Light Dependant Resistor" the amplifier circuit can be used to detect either low or high levels of temperature or light as the output voltage becomes a linear function of the changes in the active leg of the resistive bridge and this is shown below.

As Light Activated Switch

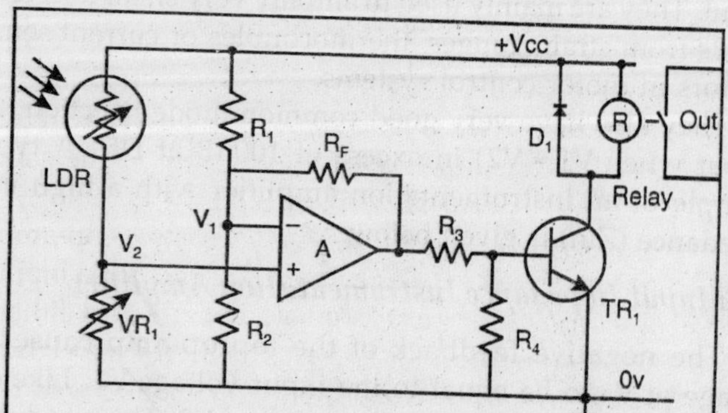

Here the circuit above acts as a light-activated switch which turns the output relay either "ON" or "OFF" as the light level detected by the LDR resistor exceeds or falls below the pre-set value of VR1.

The fixed voltage reference is applied to the inverting input terminal V1 via the R1 - R2 voltage divider network and the variable voltage (proportional to the light level) applied to the non-inverting input terminal V2.

It is also possible to detect temperature using this type of circuit by simply replacing the Light Dependant Resistor (LDR) with a thermistor. One major limitation of this type of amplifier design is that its input impedances are lower compared to that of other operational amplifier configurations, for example, a non-inverting (single-ended input) amplifier.

Each input voltage source has to drive current through an input resistance, which has less overall impedance than that of the op-amps input alone.

One way to overcome this problem is to add a *Unity Gain Buffer Amplifier* such as the voltage follower seen in the previous tutorial to each input resistor. This then gives us a differential amplifier circuit with very high input impedance and is the basis for most "Instrumentation Amplifiers".

As Instrumentation Amplifier

Instrumentation Amplifiers are high gain differential amplifiers with high input impedance and a single ended output. They are mainly used to amplify very small differential signals from strain gauges, thermocouples or current sensing resistors in motor control systems.

They also have very good common mode rejection (zero output when V1 = V2) in excess of 100dB at DC. A typical example of an instrumentation amplifier with a high input impedance (Zin) is given below:

High Input Impedance Instrumentation Amplifier

The negative feedback of the top op-amp causes the voltage at Va to be equal to the input voltage V1. Likewise, the voltage at Vb is equal to the value of V2. This produces a voltage drop across R1 which is equal to the voltage difference between V1 and V2.

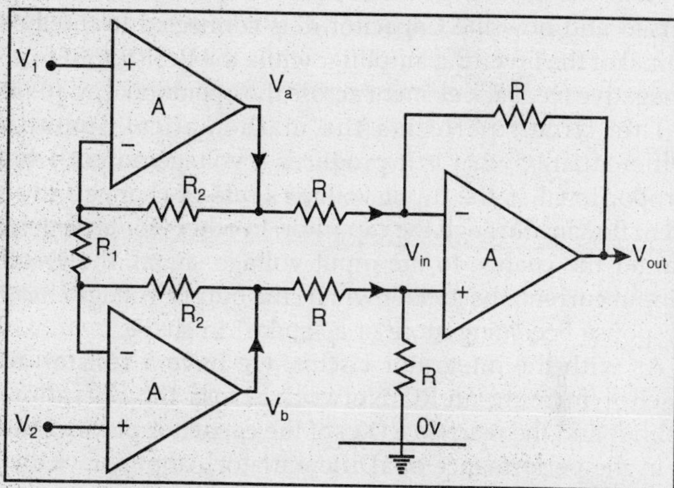

This voltage drop causes a current to flow throughR1, and as the two inputs of the buffer op-amps draw no current (virtual earth), the same amount of current flowing through R1 must also be flowing through the two resistors R2.

This then produces a voltage drop between points Va and Vb equal to:

$$\frac{(V_a - V_b)}{(V_1 - V_2)} = \left[1 + \frac{2R_2}{R_1}\right]$$

This voltage drop between points Va and Vb is connected to the inputs of the differential amplifier which amplifies it by a gain of 1 (assuming that all the "R" resistors are of equal value). Then we have a general expression for overall voltage gain of the instrumentation amplifier circuit as:

$$V_{out} = (V_1 - V_2)\left[1 + \frac{2R_2}{R_1}\right]$$

The differential gain of the circuit can be changed by changing the value of R1.

As Differentiator

The basic Differentiator Amplifier circuit is a the exact

opposite to that of the Integrator operational amplifier circuit that we saw in the previous tutorial.

Here, the position of the capacitor and resistor have been reversed and now the Capacitor, C is connected to the input terminal of the inverting amplifier while the Resistor, Rf forms the negative feedback element across the operational amplifier.

This circuit performs the mathematical operation of Differentiation, that is it produces a voltage output which is proportional to the input voltage's rate-of-change and the current flowing through the capacitor. In other words the faster or larger the change to the input voltage signal, the greater the input current, the greater will be the output voltage change in response becoming more of a "spike" in shape.

As with the integrator circuit, we have a resistor and capacitor forming an RC Network across the operational amplifier and the reactance (Xc) of the capacitor plays a major role in the performance of aDifferentiator Amplifier.

Differentiator Amplifier Circuit

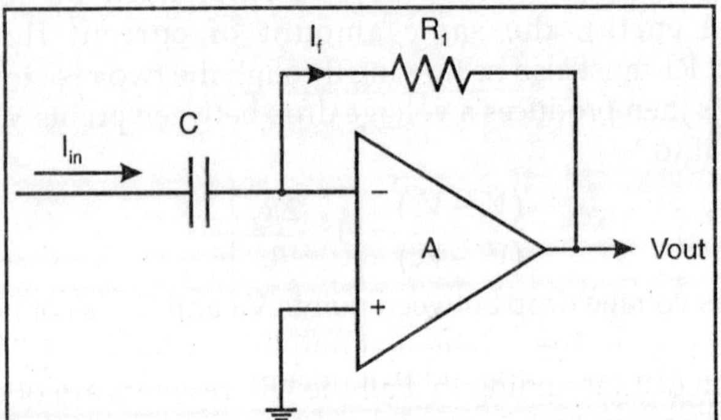

The capacitor blocks any DC content only allowing AC type signals to pass through and whose frequency is dependant on the rate of change of the input signal.

At low frequencies the reactance of the capacitor is "High" resulting in a low gain (Rf/Xc) and low output voltage from the op-amp. At higher frequencies the reactance of the capacitor is much lower resulting in a higher gain and higher output voltage from the differentiator amplifier.

However, at high frequencies a differentiator circuit becomes unstable and will start to oscillate. This is due mainly to the First-order effect, which determines the frequency response of the op-amp circuit causing a Second-order response which, at high frequencies gives an output voltage far higher than what was expected.

To avoid this the high frequency gain of the circuit needs to be reduced by adding an additional small value capacitor across the feedback resistor Rf. Ok, some math's to explain what's going on. Since the node voltage of the operational amplifier at its inverting input terminal is zero, the current, i flowing through the capacitor will be given as:

$$i_{IN} = I_F \; and = -\frac{V_{OUT}}{R_F}$$

The Charge on the Capacitor = Capacitance x Voltage across the Capacitor,

$$Q = C \times V_{IN}$$

The rate of change of this charge is,

$$\frac{dQ}{dt} = C\frac{dV_{IN}}{dt}$$

but dQ/dt is the capacitor current i

$$i_{IN} = C\frac{dV_{IN}}{dt} = I_F$$

$\therefore$

$$-\frac{V_{OUT}}{R_F} = C\frac{dV_{IN}}{dt}$$

From which we have an ideal voltage output for the Differentiator Amplifier is given as:

$$V_{OUT} = -R_F C\frac{dV_{IN}}{dt}$$

Therefore, the output voltage Vout is a constant -Rf.C times the derivative of the input voltage Vin with respect to time. The minus sign indicates a 180° phase shift because the input signal is connected to the inverting input terminal of the operational amplifier. One final point to mention, the Differentiator Amplifier circuit in its basic form has two main disadvantages compared to the previous Integrator circuit. One is that it suffers from instability at high frequencies as mentioned above, and the other is that the capacitive input makes it very susceptible to random noise signals and any noise or harmonics present in the circuit will be amplified more than the input signal itself. This is because the output is proportional to the slope of the input voltage so some means of limiting the bandwidth in order to achieve closed-loop stability is required

Differentiator Waveforms

If we apply a constantly changing signal such as a Square-wave, Triangular or Sine-wave type signal to the input of a differentiator amplifier circuit the resultant output signal will be changed and whose final shape is dependant upon the RC time constant of the Resistor/Capacitor combination.

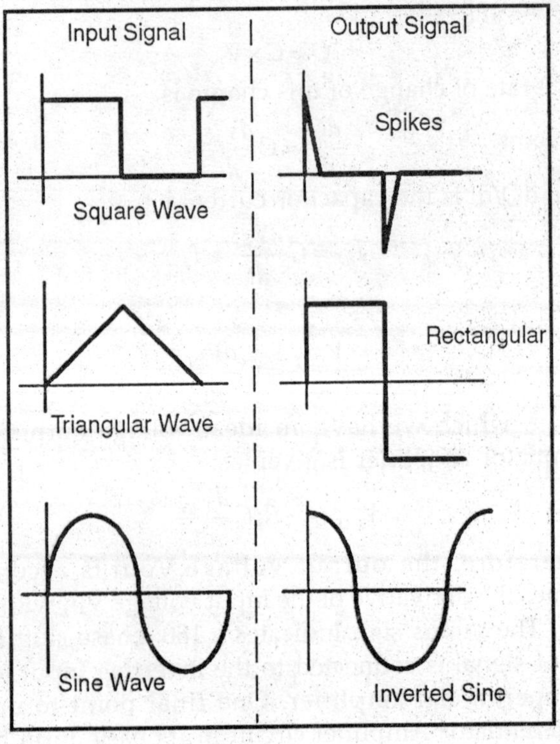

Improved Differentiator Amplifier

The basic single resistor and single capacitor differentiator circuit is not widely used to reform the mathematical function of Differentiation because of the two inherent faults mentioned above, Instability and Noise.

So in order to reduce the overall closed-loop gain of the circuit at high frequencies, an extra Resistor, R2 is added to the input.

Improved Differentiator Amplifier Circuit

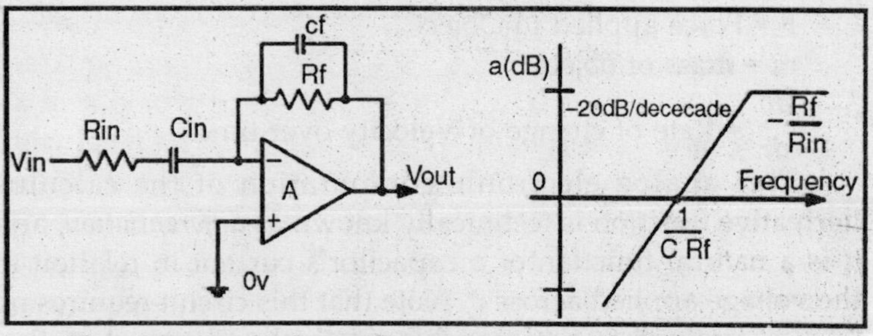

The circuit which we have now acts like a Differentiator amplifier at low frequencies and an amplifier with resistive feedback at high frequencies giving much better noise rejection.

LARGE SIGNAL OPERATION OF OP-AMP

The reactive properties of capacitors and inductors lend themselves well to the simulation of variables related by calculus functions.

Remember how the current through a capacitor was a function of the voltage's rate of change, and how that rate of change was designated in calculus as the *derivative*?

Well, if voltage across a capacitor were made to represent the velocity of an object, the current through the capacitor would represent the force required to accelerate or decelerate that object, the capacitor's capacitance representing the object's mass:

$$i_C = C\frac{dv}{dt}$$

Where,

$$i_C = C\frac{dv}{dt}$$

i_C = Instantaneous Current through capacitor.
C = Capacitance in farads
$\frac{dv}{dt}$ = Rate of Change of Voltage Over time

$$F = m\frac{dv}{dt} =$$

where,
F = Force applied to Object
m = mass of object
$\frac{dv}{dt}$ = Rate of change of Velocity over time

This analog electronic computation of the calculus derivative function is technically known as *differentiation,* and it is a natural function of a capacitor's current in relation to the voltage applied across it. Note that this circuit requires no "programming" to perform this relatively advanced mathematical function as a digital computer would.

Electronic circuits are very easy and inexpensive to create compared to complex physical systems, so this kind of analog electronic simulation was widely used in the research and development of mechanical systems. For realistic simulation, though, amplifier circuits of high accuracy and easy configurability were needed in these early computers.

It was found in the course of analog computer design that differential amplifiers with extremely high voltage gains met these requirements of accuracy and configurability better than single-ended amplifiers with custom-designed gains.

Using simple components connected to the inputs and output of the high-gain differential amplifier, virtually any gain and any function could be obtained from the circuit, overall, without adjusting or modifying the internal circuitry of the amplifier itself.

These high-gain differential amplifiers came to be known as *operational amplifiers,* or *op-amps,* because of their application in analog computers' mathe-matical *operations.*

Modern op-amps, like the popular model 741, are high-performance, inexpensive integrated circuits. Their input impedances are quite high, the inputs drawing currents in the range of half a microamp (maximum) for the 741, and far less for op-amps utilizing field-effect input transistors.

Output impedance is typically quite low, about 75 Ω for the model 741, and many models have built-in output short circuit protection, meaning that their outputs can be directly shorted to ground without causing harm to the internal circuitry.

With direct coupling between op-amps' internal transistor stages, they can amplify DC signals just as well as AC (up to certain maximum voltage-risetime limits).

It would cost far more in money and time to design a comparable discrete-transistor amplifier circuit to match that kind of performance, unless high power capability was required. For these reasons, op-amps have all but obsoleted discrete-transistor signal amplifiers in many applications.

The following diagram shows the pin connections for single op-amps (741 included) when housed in an 8-pin DIP (Dual Inline Package) integrated circuit:

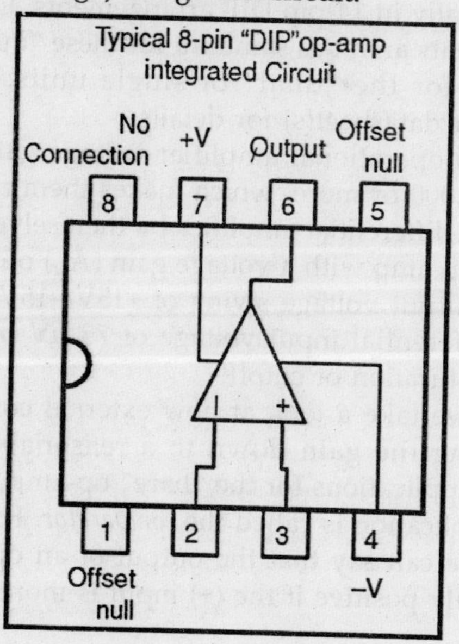

Some models of op-amp come two to a package, including the popular models TL082 and 1458. These are called "dual" units, and are typically housed in an 8-pin DIP package as well, with the following pin connections:

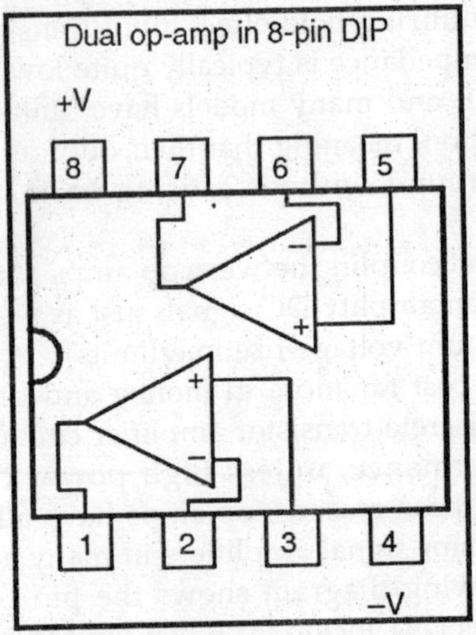

Operational amplifiers are also available four to a package, usually in 14-pin DIP arrangements. Unfortunately, pin assignments aren't as standard for these "quad" op-amps as they are for the "dual" or single units. Consult the manufacturer datasheet(s) for details.

Practical operational amplifier voltage gains are in the range of 200,000 or more, which makes them almost useless as an analog differential amplifier by themselves.

For an op-amp with a voltage gain (A_V) of 200,000 and a maximum output voltage swing of +15V/–15V, all it would take is a differential input voltage of 75 μV (microvolts) to drive it to saturation or cutoff!

Before we take a look at how external components are used to bring the gain down to a reasonable level, let's investigate applications for the "bare" op-amp by itself.

One application is called the *comparator*. For all practical purposes, we can say that the output of an op-amp will be saturated fully positive if the (+) input is more positive than the (–) input, and saturated fully negative if the (+) input is less positive than the (–) input.

In other words, an op-amp's extremely high voltage gain makes it useful as a device to compare two voltages and change output voltage states when one input exceeds the other in magnitude.

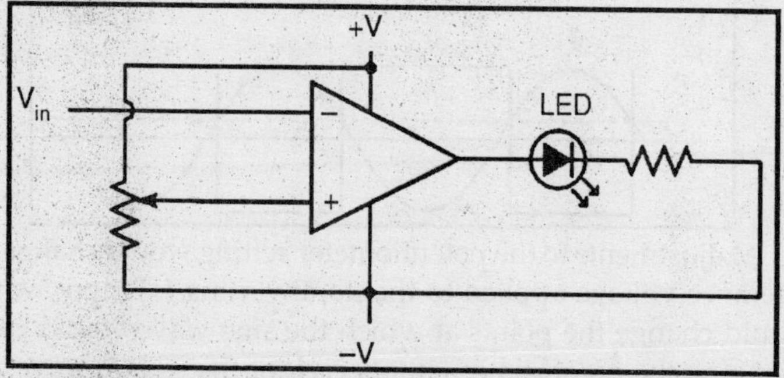

In the above circuit, we have an op-amp connected as a comparator, comparing the input voltage with a reference voltage set by the potentiometer (R_1).

If V_{in} drops below the voltage set by R_1, the op-amp's output will saturate to +V, thereby lighting up the LED. Otherwise, if V_{in} is above the reference voltage, the LED will remain off.

If V_{in} is a voltage signal produced by a measuring instrument, this comparator circuit could function as a "low" alarm, with the trip-point set by R_1.

Instead of an LED, the op-amp output could drive a relay, a transistor, an SCR, or any other device capable of switching power to a load such as a solenoid valve, to take action in the event of a low alarm.

Another application for the comparator circuit shown is a square-wave converter. Suppose that the input voltage applied to the inverting (–) input was an AC sine wave rather than a stable DC voltage.

In that case, the output voltage would transition between opposing states of saturation whenever the input voltage was equal to the reference voltage produced by the potentiometer.

The result would be a square wave:

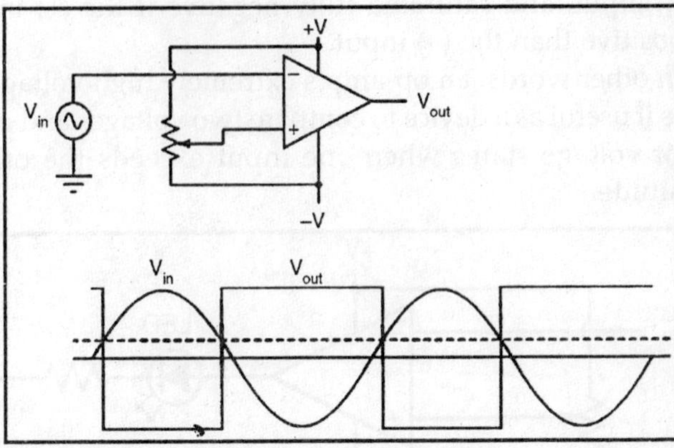

Adjustments to the potentiometer setting would change the reference voltage applied to the noninverting (+) input, which would change the points at which the sine wave would cross, changing the on/off times, or*duty cycle* of the square wave:

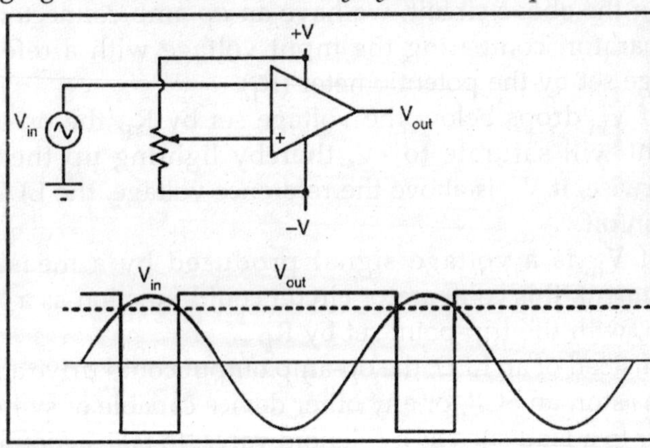

It should be evident that the AC input voltage would not have to be a sine wave in particular for this circuit to perform the same function. The input voltage could be a triangle wave, sawtooth wave, or any other sort of wave that ramped smoothly from positive to negative to positive again. This sort of comparator circuit is very useful for creating square waves of varying duty cycle. This technique is sometimes referred to as*pulse-width modulation,* or PWM (varying, or *modulating* a waveform according to a controlling signal, in this case the signal produced by the potentiometer).

Another comparator application is that of the bargraph driver. If we had several op-amps connected as comparators, each with its own reference voltage connected to the inverting input, but each one monitoring the same voltage signal on their noninverting inputs, we could build a bargraph-style meter such as what is commonly seen on the face of stereo tuners and graphic equalizers. As the signal voltage (representing radio signal strength or audio sound level) increased, each comparator would "turn on" in sequence and send power to its respective LED. With each comparator switching "on" at a different level of audio sound, the number of LED's illuminated would indicate how strong the signal was.

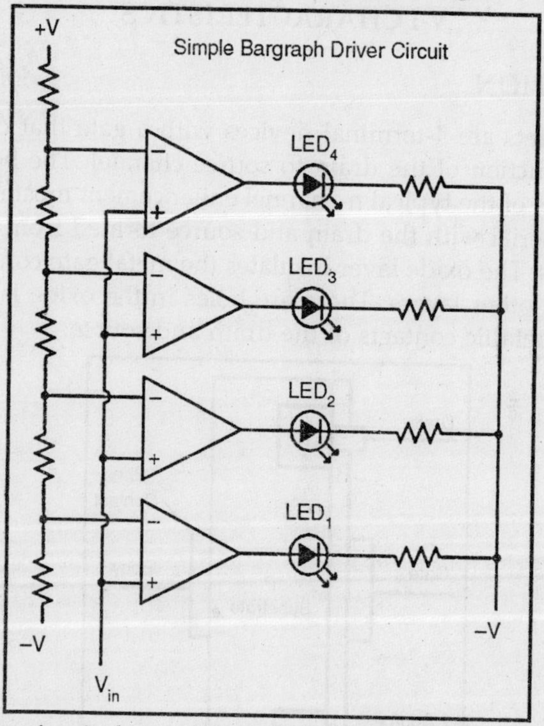

In the circuit shown above, LED_1 would be the first to light up as the input voltage increased in a positive direction. As the input voltage continued to increase, the other LED's would illuminate in succession, until all were lit.

Chapter 7

Mosfet

STRUCTURE OPERATION AND V-I CHARACTERISTICS

DEFINITION

Mosfets are 4-terminal devices with a gate that controls the conduction of the drain to source channel. The body (or substrate) of the typical n-channel enhancement mosfet is a P-type material with the drain and source formed from N-type materials. The oxide layer insulates the metal gate connection from the other layers. There are holes in the oxide layer left for the metallic contacts of the drain and source.

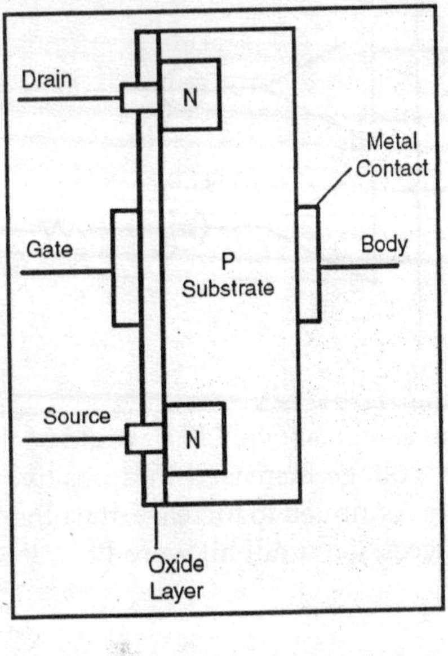

STRUCTURE

N-P junctions act as diodes, and because the layers of the mosfet have P-type and N-type materials in close contact, diodes are formed from the body to the drain or source, as shown in the drawing here of the 4-terminal mosfet.

If you were to leave the body connection floating, there could be no voltage flow through the two PN body diodes because they are reverse connected to each other.

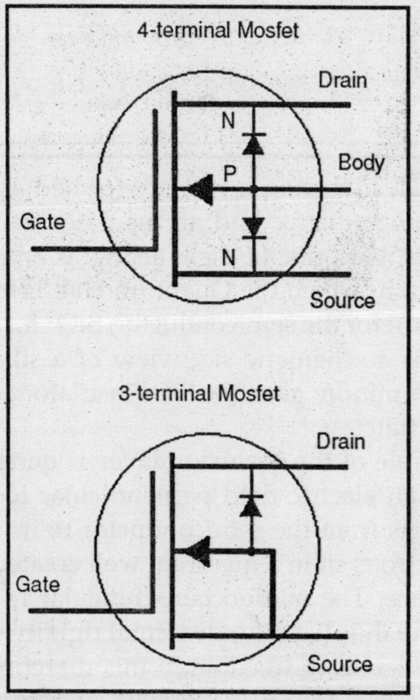

However, in the 3-terminal mosfets (like the BS170) that are commonly used in guitar effects pedals, the body and source are connected internally and this shorts out the PN junction of the lower diode and removes it from the circuit.

The top body diode can now conduct from drain to source because of the internal connection and therefore has to be taken into consideration during circuit design. This means that every 3-terminal mosfet has a body diode due the construction of the device. It is not added as a separate component inside the mosfet.

WORKING

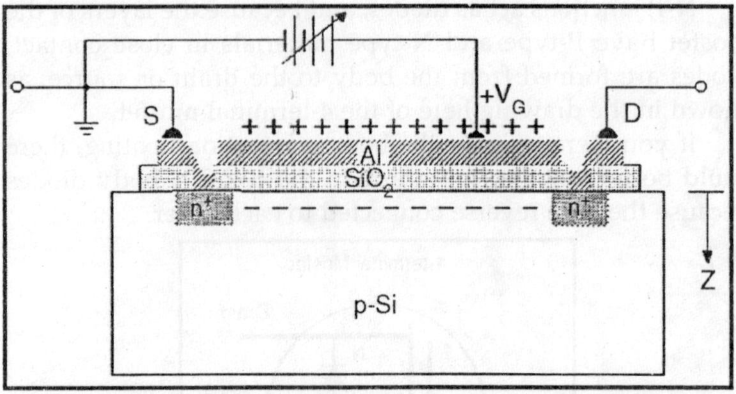

In MOSFET, inversion layers are formed at the interface between a semiconductor and an insulator or between two semiconductors, with one of them acting as an insulator.

The system in which the Quantum Hall Effect(QHE) was discovered has Si for the semiconductor, SiO^2 for the insulator. Figure above is a schematic side view of a silicon MOSFET showing the aluminum gate, the SiO^2 insulator and the p-type Si crystal substrate.

The principle of the inversion layer is quite simple. It is arranged that an electric field perpendicular to the interface attracts electrons from the semiconductor to it.

These electrons sit in a quantum well created by this field and the interface. The motion perpendicular to the interface is quantized and thus has a fundamental rigidity which freezes out motional degrees of freedom in this direction.

The result is a two-dimensional system of electrons. Further, the wavelengths of these electrons are long so that an effective mass approxiamtion with parabolic bands is quite good. The total self-consistent potential seen by the electrons is conveniently described by the picture of "band bending".

That is to say, the periodic lattice potential gives rise to energy bands, and the slowly varying electric potential then is regarded as bending these bands. Figure gives the schematic diagram for this process.(Charge density $\sim 10^{13} cm^{-2}$)

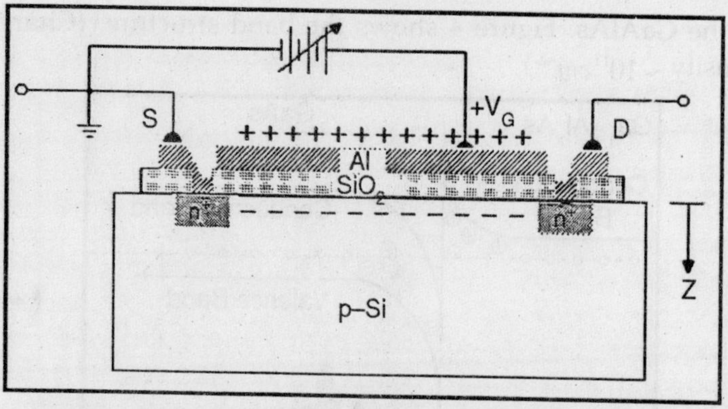

Fig. Electron Energy Level Diagram of a Silicon MOSFET with a Positive Voltage Applied to the Aluminum Gate

Another type of two-dimensional electron system is formed in the heterostructures of two semiconductors. Using molecular beam epitaxy(MBE) technique, people can grow two semiconductors alternately to form a one dimensional sandwich like strucrure.

Each layer has a width of about several nanometers. They are called superlattice. They can also be grown by metal-organic chemical vapour deposition(MOCVD). For example, in GaAs-Ga $_{1-x}Al_xAs$ superlattice, a certain controlled number of layers of GaAs is followed by an almost perfectly matched sequence of layers of GaAlAs.

The GaAlAs is deliberately doped n-type, which puts mobile electrons into its conduction band. These electrons will migrate to fill the few holes on the top of the GaAs valence band but most of them will end up in states near the bottom of the GaAs conduction band. However, there is a positive charge left on the donor impurities which attracts these electrons to the interface and bends the bands in the process.

This is the source of the electric field in this system. The transfer of electrons from GaAlAs to GaAs will continue until the dipole layer formed from the positive donors and the negative inversion layer is sufficiently strong.

This dipole layer gives rise to a potential discontinuity which finally makes the Fermi level of the GaAs equal to that of the GaAlAs. Figure 4 shows the band structure. (Charge density $\sim 10^{11} cm^{-2}$)

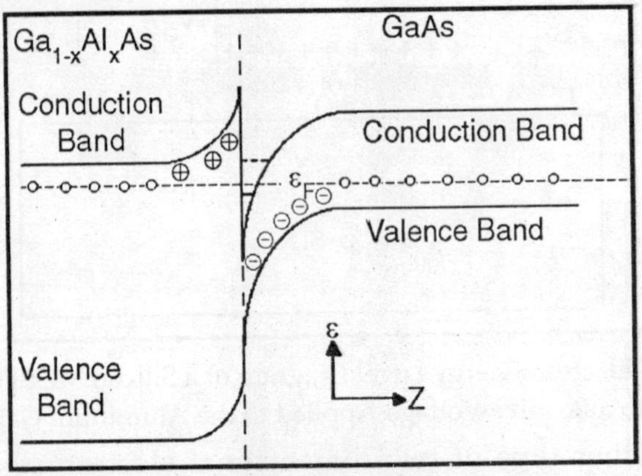

Fig. Electron Energy Level Diagram of a GaAs-AlGaAs Heterostructure Device

Two-dimensional electron system can also be formed in the surface of liquid Helium. There exists a potential barrier of about 1eV in the surface of liquid Helium which prevents electrons from transmitting into the liquid.

On the other hand, the mirror potential attracts the electrons in the surface, resulting a 2D electron system.(Charge density $\sim 10^{9} cm^{-2}$)

WORKING

As can be seen on the figure the source and drain regions are identical. It is the applied voltages which determine which n-type region provides the electrons and becomes the source, while the other n-type region collects the electrons and becomes the drain.

The voltages applied to the drain and gate electrode as well as to the substrate by means of a back contact are refered to the source potential, as also indicated on the figure.

A top view of the same MOSFET is shown in Fig., where the gate length, *L*, and gate width, *W*, are identified. Note that the gate length does not equal the physical dimension of the gate, but rather the distance between the source and drain regions underneath the gate. The overlap between the gate and the source and drain region is required to ensure that the inversion layer forms a continuous conducting path between the source and drain region.

Typically this overlap is made as small as possible in order to minimize its parasitic capacitance.

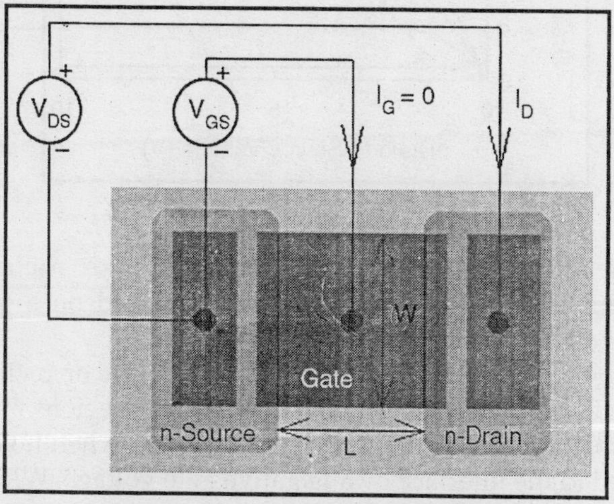

Fig. View of an n-type Metal-Oxide-Semiconductor-Field-Effect-Transistor

The flow of electrons from the source to the drain is controlled by the voltage applied to the gate. A positive voltage applied to the gate, attracts electrons to the interface between the gate dielectric and the semiconductor.

These electrons form a conducting channel between the source and the drain, called the inversion layer. No gate current is required to maintain the inversion layer at the interface since the gate oxide blocks any carrier flow. The net result is that the current between drain and source is controlled by the voltage which is applied to the gate.

V-I CHARACTERISTICS

The typical current versus voltage (I-V) characteristics of a MOSFET are shown in the figure below. Implemented is the quadratic model for the MOSFET.

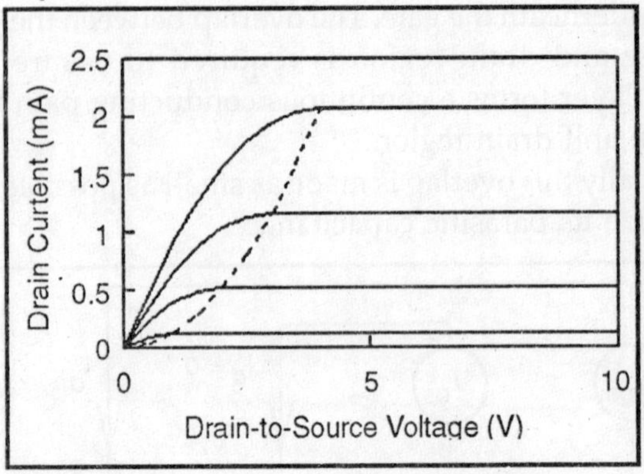

Fig. *I-V Characteristics of an n-type MOSFET*

We will primarily discuss the *n-type* or n-channel MOSFET. This type of MOSFET is fabricated on a p-type semiconductor substrate.

The complementary MOSFET is the *p-type* or p-channel MOSFET. It contains p-type source and drain regions in an n-type substrate. The inversion layer is formed when holes are attracted to the interface by a negative gate voltage. While the holes still flow from source to drain, they result in a negative drain current. CMOS circuits require both n-type and p-type devices.

MOSFET AS AMPLIFIER AND SWITCH

MOSFET AS AN AMPLIFIER

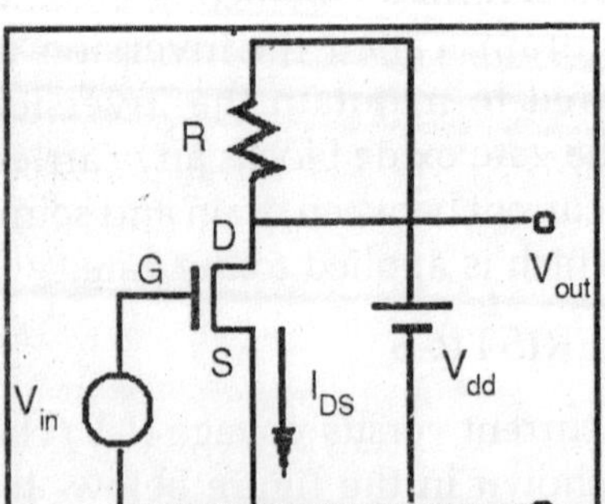

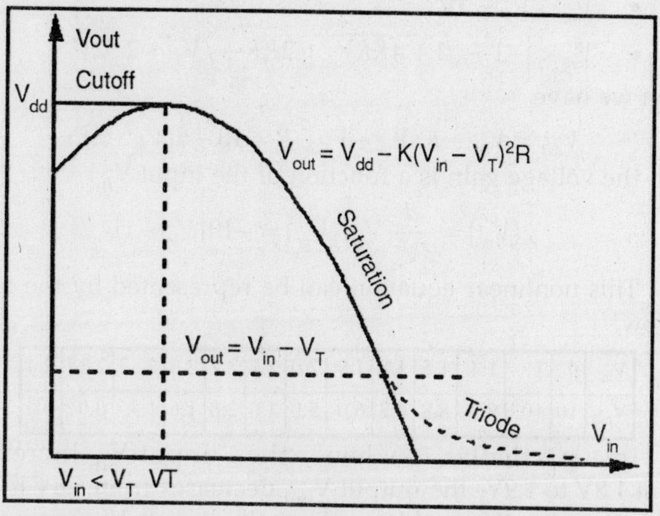

Assume in the circuit above $V_{in} = V_{GS} > V_T$ and the transistor is in saturation region, i.e., $V_{DS} > V_{in} - V_T$, then we have,

$$\begin{cases} I_{DS} = K(V_{in} - V_T)^2 \\ V_{\infty\varepsilon} = V_{DS} = V_{sd} - I_{DS}R = V_{dd} - KR(V_{in} - V_T)^2 \end{cases}$$

The second equation relates the output V_{out} to the input V_{in}, as shown by the red segment of the curve in the plot above. As the transistor is in saturation region,

$$V_{out} = V_{dd} - KR(V_{in} - V_T)^2 \geq V_{in} - V_T$$

which can be solved for V_{in} to get:

$$V_{in} < \frac{-1 + \sqrt{1 + 4KRV_{dd}}}{2KR} + V_T$$

It can be seen that when the transistor is in saturation mode the slope of the curve (red) indicates the ratio between input V_{in} and output V_{out}. And the voltage gain of the circuit is defined as:

$$\vartheta = \frac{dV_{out}}{dV_{in}} = \frac{d}{dV_{in}}\left(V_{dd} - K(V_{in} - V_T)^2 R\right) = -2KR(V_{in} - V_T)$$

Example: Assume V_{dd} =10V, R=10kΩ, K = 0.5 mA/V^2, V_T = 1V. For the transistor to be in saturation region, we need

- $V_{in} > V_T = 1V$
- $V_{in} < \left|-1+\sqrt{1+4KRV_{dd}}\right|/2KR+/V_T = 2.32V$

then we have

$$V_{out} = V_{dd} - K(V_{in} - V_T)^2 R = 10 - 5(V_{in} - 1)^2$$

and the voltage gain is a function of the input V_{in}:

$$g(V_{in}) = \frac{d}{dV_{in}} V_{out}(V_{in}) = -10(V_{in} - 1)$$

This nonlinear equation can be represented by the table below:

V_{in}	0	1	1.4	1.5	1.8	1.9	2.0	2.1	2.2	2.3	2.32	2.35	2.4
V_{out}	10	10	9.2	8.8	6.8	6.0	5.0	4.0	2.8	1.6	1.3	0.9	0.0

In particular, when the input V_{in} increases from 1.8V to 1.9V, the output V_{out} decreases from 6.8V to 6V, with a gaing(V_{in}). Also when the input V_{in} increases from 2.0V to 2.1V, the output V_{out} decreases from 5V to 4V, with a gain $g(V_{in}) = g(2) = -10(2-1) = -10$.

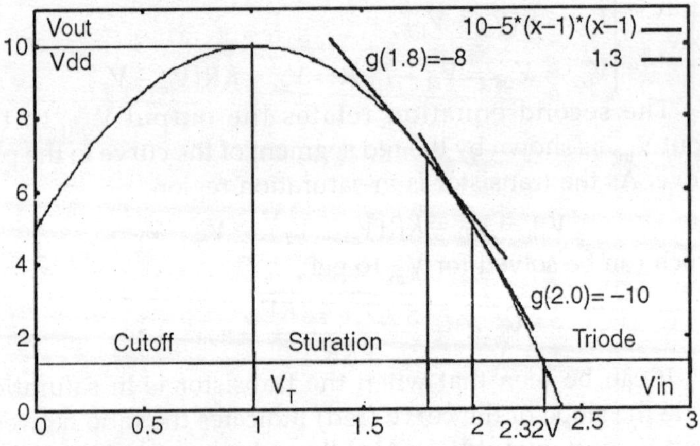

In summary, we see that

- When the transistor is in saturation mode, the circuit behaves as a voltage amplifier.
- The voltage gain is the slope of the tangent of the curve (red) as a function of V_{in}.
- The value of the gain depends on the level of

input V_{in}. When $V_T = 1V < V_{in} < 2.32V$, the gain $|g| > 1$ is greater than one.

- The output voltage is 180° out of phase with the input voltage (g<0), as the slope of V_{out} V_{in} is negative.
- When V_{in} <V_T =1V, the transistor is cutoff. On the other hand, when $V_{in} > 2.32V$ V_{out} is more than one V_T below V_{in}, for example, $V_{in} = 2.32, V_{out} = 1.3 < V_{in} - V_T = 2.32 - 1 = 1.32$, the transistor is in triode region. In either of the two cases, the transistor has no amplification capability.

Next we consider a MOSFED circuit with sinusoidal input. Assume the drain resistor is $R = 10k\Omega$, K=1mA/V^2, V_T = 1V, V_{dd}=5V and a sinusoidal input $U_{in}(t)$ = $V_{\varepsilon i}$+sin(ωt). If the bias voltage is V_{out} =1.5V, the input voltage $u_{in}(t)$ varies between 1.4V and 1.6V.

The output voltage can be found to be:

$$v_{out} = V_{dd} - RI_{DS} = V_{dd} - RK(V_{in} - V_T)^2 = 5 - 10(V_{in} - 1)^2$$

In particular, corresponding to $V_{in} = 1.4V, 1.5V, 1.6V$, the output voltage u_{out} and the current i_{DS} are, respectively, u_{out} = 3.4V, 2.5, 1.4V, and i_{DS} =0.16mA, 0.25mA, 0.36mA, as shown in the figure below:

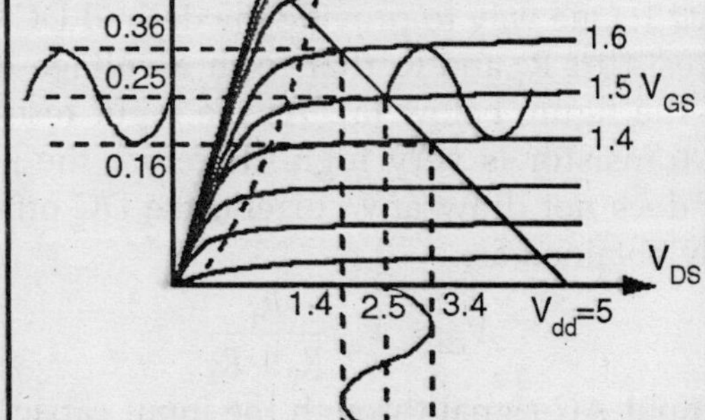

Biasing: In the example above, the DC offset of the input is at 1.5V, so that the transistor is working in the saturation region when the magnitude of the AC input is limited. However, if this offset is either too high or too low, the gate voltage may go beyond the saturation region to enter either the triode or the cutoff region. In either case, the output voltage will be severely distorted, as shown below:

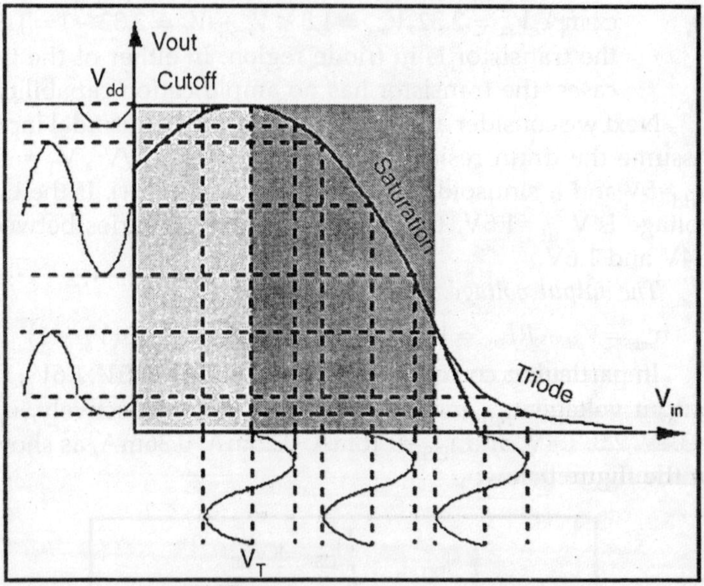

It is therefore clear that the DC offset or biasing gate voltage has to be properly setup to make sure the dynamic range of the input signal is within the saturation region.

Method 1: One way to provide the desired DC offset is to use two resistors R_1 and R_2 that form a voltage divider, as shown in the figure below (a). As the input resistance of a MOSFET transistor is very high, therefore the gate of the transistor does not draw any current, the DC offset voltage can simply obtained as:

$$V_{out} = V_{dd} \frac{R_1}{R_1 + R_2}$$

The input AC signal through the input capacitor is then superimposed on this DC offset.

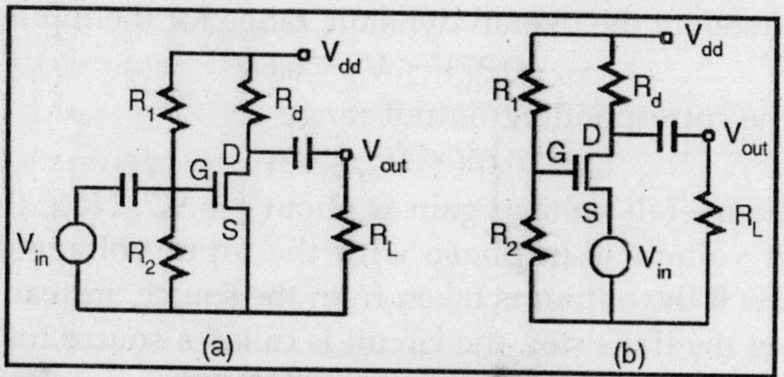

Method 2: Another way to set up the bias is the circuit shown in (b) above. Assume $R_1 = 84k$, $R_2 = 16k$, $V_{dd} = 10V$, and $K = 0.5mA/V^2$. The bias voltage can be found to be $V_B = V_{out} = 1.6V$, and the voltage between gate and source is $V_{GS} = V_B - V_{in}$.

The output voltage is:

$$V_{out} = V_{dd} - R_d K(V_{GS} - V_T)^2 = V_{dd} - R_d K(V_B - V_{in} - V_T)^2$$

$$= 10 - 10 \times (0.6 - V_{in})^2$$

When $V_{in} = 0$, $V_{out} = 6.4V$.

To determine the dynamic range of the input V_{in}, recall the conditions for the transistor to be in saturation region:

- To avoid cutoff region: $V_{GS} - V_T \geq 0$. For this particular circuit,

$$V_{GS} - V_T - V_B - V_{in} - V_T = 0.6 - V_{in} \geq 0$$

Solving this we get $V_{in} \leq 0.6V$ with corresponding output $V_{out} \leq 10V$.

- To avoid triode region: $V_{DS} \geq V_{GS} - V_T$. For this particular circuit,

$$V_{DS} = V_{out} - V_{in} = (V_{dd} - R_d I_{DS}) - V_{in} = V_{dd} - R_d K$$

$$(V_B - V_{in} - V_T)^2 - V_{in} \geq V_B - V_{in} - V_T.$$

that is,

$$V_{out} = V_{dd} - R_d K(V_B - V_{in} - V_T)^2 \geq V_B - V_T$$

i.e.,

$$V_{out} = 10 - 10 \times (0.6 - V_{in})^2 \geq 0.6$$

Solving this for V_{in} we get $V_{in} \geq -0.37V$, with corresponding output $V_{out} \geq 0.6V$

Therefore the overall dynamic range for the input is,

$$-0.37V \le V_{in} \le 0.6V$$

with the corresponding output range

$$0.6V \le V_{out} \le 10V$$

and the overall voltage gain is about g = 9.7. Note that the output voltage is in phase with the input voltage. *Source Follower*: If the output is taken from the source, instead of the drain of the transistor, the circuit is called a source follower.

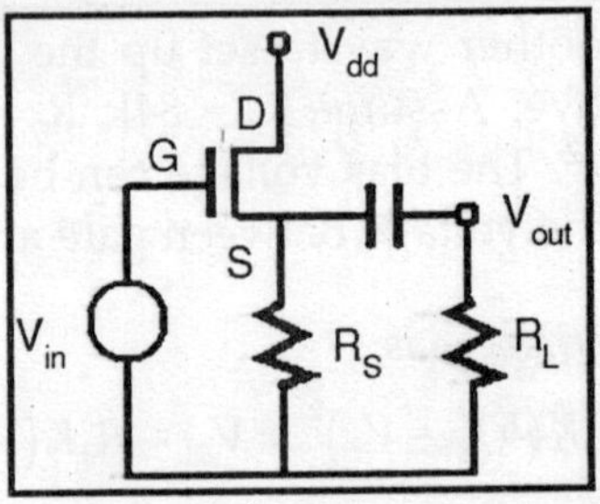

Assume $R_S = 10k\Omega$, $V_T = 1V$ and $K = 10mA/V^2$. To find the input and output voltages and the gain of the circuit, consider the current $I_{DS} = V_{out}/R_S$:

$$I_{DS} = K(V_{GS} - V_T)^2 = K(V_{in} - V_{out} - V_T)^2 = V_{out}/R_S$$

Plugging in the given values, we get,

$$(V_{in} - V_{out} - 1)^2 = V_{out}$$

If V_{in} *= 2V, this equation becomes:*

$$(V_{out} - 1)^2 = V_{out}$$

which can be solved to get V_{out} = 2.6V or V_{out} = 0.4V. We take the smaller voltage in order for the transistor to be outside the cutoff region:

$$V_{GS} = V_{in} - V_{out} = 2 - V_{out} > V_T = 1$$

Similarly, if V_{in} = 3V, the equation becomes:.

$$(V_{out} - 2)^2 = V_{out}$$

and we get V_{out} = 1V. The voltage gain of the source follower is

$$g = \frac{\Delta V_{out}}{\Delta V_{in}} = \frac{1 - 0.4}{3 - 2} = 0.6 < 1$$

MOSFET AS AN SWITCH

- There are several types of Metal Oxide Semiconductor Field-Effect Transistors, also known as MOSFETs. MOSFETs are devices used as transistors in analog or digital circuits. They can either switch or amplify electronic signals. Power MOSFETs are specifically designed to manage large amounts of power and have high impedance levels.

- Power MOSFETs were first developed in the late 1970's, first being created by International Rectifier in 1978. While Power MOSFETs perform the same functions as Bi-polar transistors, Power MOSFETs are voltage controlled, while Bi-polar transistors are current controlled.
- Power MOSFETs are efficient at lower voltages, and are the most common low-voltage switches. They can be found in most power converters, power supplies and low-voltage controllers. Power MOSFETs utilize a vertical structure, which allows the transistor to sustain high current with high-blocking voltage.
- MOSFETs operate under the application of power from a power source. When positive voltage comes from a power source, electrons are attracted to the positive voltage. If the voltage is above a certain level of voltage, enough electrons will be attracted and will accumulate to form a conducting channel across the body of the MOSFET. This channel allows electrons to flow in either direction.

BIASING IN MOS AMPLIFIER CIRCUITS

Biasing is the static voltage and current settings for a given device for certain functions of operation. Depending on the needs of the circuit, there may be as many as six different biasing schemes. But for the most part there are four basic biasing schemes.

They are Class A, Class B, Class C and Class AB.

Class A: This is where the device is biased at a point where it will conduct and amplify for the entire signal.

Class B: This is where the device is biased such that it will conduct for only half of the signal.

Class C: This is where the device will only conduct for a small portion of the peak of the signal. It can also be described as any biasing point where the device conducts for less than half of the signal. This biasing scheme is never used in audio and can be found in R.F. applications.

Class AB: This is where the device conducts for more than half of the signal but less than the full signal.

This is as far as I will go here regarding biasing classes. Here we will discuss practical implementation of the different biasing schemes.

With tube amplifiers, biasing has been the subject of many many articles and discussions on news group threads. It also seems to be one of the least understood subjects with regard tube sound.

BIAS SETTING SOUND "BETTER" THAN ANOTHER

First, let us examine a simple single stage triode circuit with a standard bias and bypass capacitor:

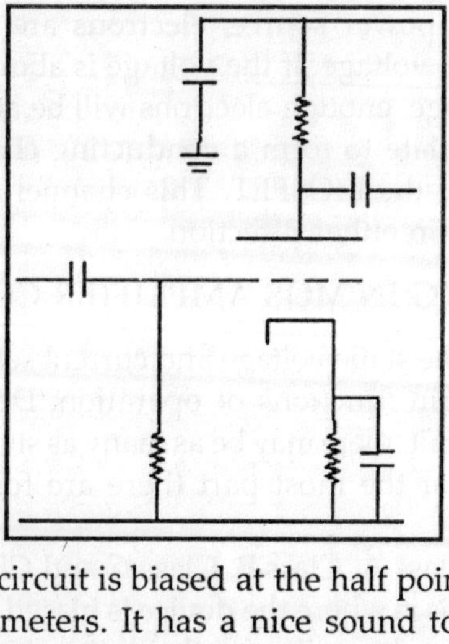

The above circuit is biased at the half points of the tube's maximum parameters. It has a nice sound to it and is loud but has some noise (hiss).

However, at strong loud bassy passages, it tends to seem limited but warmer. Sometimes it even seems to drop out. Why is this? Some would claim insufficient transient response from the power supply. It sounds slow, as many would say.

But let's take a closer look at this circuit. Let us rotate the circuit 90 degrees to the left. Now let's remove the grid circuit for a minute.

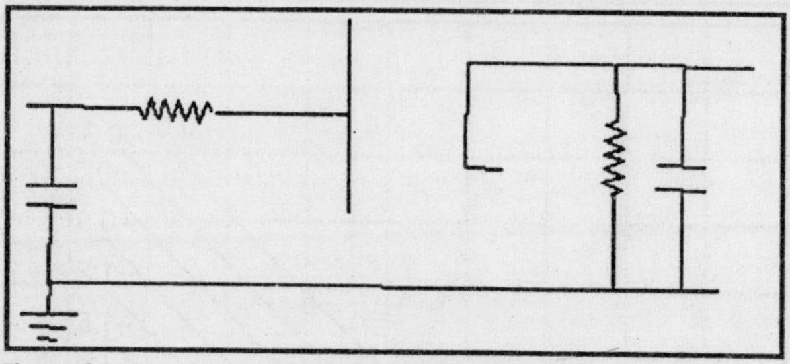

What does it look like now? That's right, a power supply rectifier! A half wave rectifier to be exact. So now what do the cathode resistor and bypass cap look like? A load and *filter*. The bypass cap sees a constant DC signal accompanied by an AC component. To a filter cap, or any capacitor for that matter, this appears as a fluctuating DC, or power supply ripple!

To get a bit deeper into the physics of the thing, if one were to calculate the time constant of the RC circuit at the cathode, one will find that for the above circuit the time constant is t=RC, or for the sake of argument, we will use the values I used for the actual circuit, 2200 X.000022 (22 microfarads).

The answer is.0484 seconds. What that means is that the capacitor resistor circuit requires about.0484 seconds to charge 63 per cent of its full capacity.

For the first.0484 seconds, if the capacitor were a 25 volt capacitor, and the bias were set to 2 volts, it would only take the first time constant to reach 2 volts. .0484 seconds factors out to a frequency of 20 hertz (1/t). If I took a 20 hertz signal of.7 volt RMS and inserted it, the positive peak will be 1 volt above the bias of 2 volts. Since the time constant is as small as it is, it wouldn't take much to increase the voltage of the capacitor.

This also simultaneously changes the bias of the circuit by about.4 volts (0.404 × Vpeak. The average of the positive portion of a sine wave). Looking at the curve chart below for a triode, this will change the characteristics of the tube which is then amplified through the tube. A larger capacitor will do this more and make things worse.

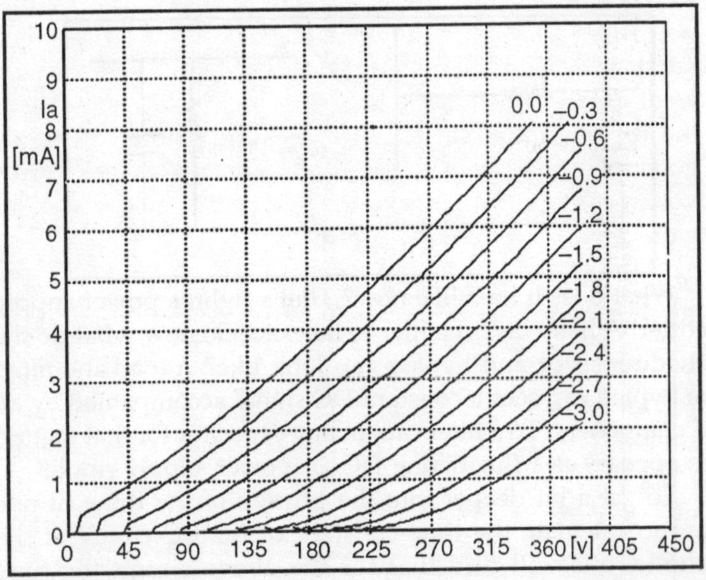

Looking at this chart, go to the 225 on the bottom line. This is the B+. Then go up to the 2.5 line from the left (the Y axis). This is the quiescent, or no input signal, current. For these conditions, it is showing a -1.4 volt grid bias. If the bias were anything different, then one would have to follow the line in the following chart to see the amount of voltage swing the input would accept before clipping.

A method for biasing a tube is to draw a line from the B+ voltage to the desired current and choose a bias setting accordingly.

This line is drawn for the modified version of the above circuit I currently listen to. The B+ is 290 volts and the current is about 2 milliamps.

If you were to place a pointer at the bias point you designed the amp for, and follow the line, you will see the varying current and the voltage it will produce. Say I had biased the tube for 1.5 volts ("I had a biased point for 1.5 volts"). No I didn't mean for *you* to say it! I start at 1.5 volts on the line, go to zero and I will see that the plate voltage has gone down to 50 volts, while the current went up to 1.8 mA or so. Then I take the voltage down to -3 volts.

This would be a full peak to peak input swing. We can see that the voltage went to about 260 volts, where the current went down to about.3 mA. So the conclusion is that we use the above graph to set static voltages and predict to a relatively good degree the gain. So the output voltage swings about 210 volts peak to peak. This makes it about 74 volts RMS. So the gain can be calculated as Vout/Vin, or 74/1.06= about 70.

Theoretically, that is:

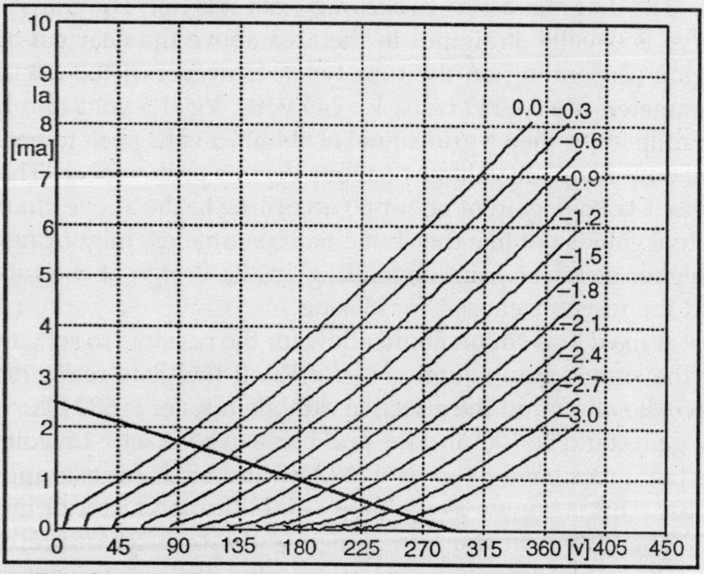

For the sound quality we want, we need to use a transfer chart in conjunction with this one. It is a chart which shows the relationship between grid voltage and plate current.

If we bias the tube along the most linear part of this line, and follow it with a nominal voltage swing across to the above chart, then one can find where clean signals end and clipping begins. Unfortunately I do not have a true chart, but the below will illustrate what I mean.

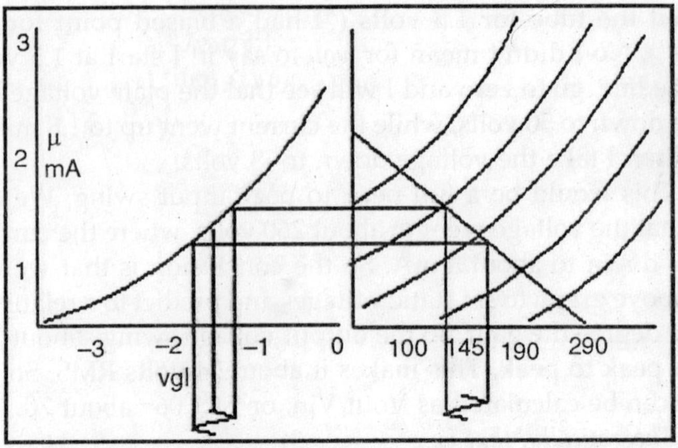

The diagram assumes accuracy, but it is not. The transfer curve is usually straighter in the area above the near cut-off region (depending on the tube type). However, if the circuit parameters are B+ 290 volts, Vp 145 volts, Vg -1.5 volts and Ip 1.5 milliamps, then a grid signal of about.75 volts peak to peak will output a plate voltage of about 45 volts peak to peak. That gives a voltage gain of about 60 according to the above chart. Actual gain is not like this, but circuit parameters might cause a higher or lower gain, depending on the design of the tube and the resistors chosen for biasing.

A fixed bias circuit could allow for the resistors to set gain by the simple gain equation G=Rp/Rk. If the plate resistance is 65 kilo ohms and the effective cathode resistor is 680 Ohms, the gain could be 100, and the grid bias can be exactly 1.5 volts. Certain tubes have a top gain (12AX7 and 6EU7 for example) of 100. This is known as the tube's amplification factor or mu.

The curve on the left, which is the transfer curve, is actually more curved as one approaches the more negative voltage level. Biasing down there will not yield a very linear response. So fixed biasing has its advantages, one being the possibility of not using a bypass resistor. Another is accurate biasing that doesn't change much with time or heat, providing better linearity. In audio, with one instrument or even the fundamental being at a high amplitude, the instant that the peak reaches this area, another signal of a higher frequency may suffer the effects of distortion. This is a form of modulation distortion.

Capacitors cause current to lag the voltage, so the bias will not follow the signal, hence altering the incoming signal's nature, and the overall sound quality. What this means is that a signal whose current component and voltage component are in phase will come out of a capacitor with current being 60 degrees behind the voltage signal.

This may not necessarily be a problem if there is no grid resistor (some circuits do this), but if there is a grid resistor, then the resistor will react to the varying current and add to the original voltage signal the resultant voltage signal across the resistor and cause a slight phase distortion.

This is because, not only do the biasing schemes mentioned at the outset of this discussion alter dramatically the sound quality, but also variations within the most popular, the class A, also alter the sound quality.

You will find that the schemes that use the lower voltages and currents are more popular because they are the least linear. This may seem like a contradiction to what an audiophile would want from a sound system, especially one that uses tube and expensive irons. It is.

Being a purist who is also a tube lover, I find this to be perplexing. In my opinion, wherever the tubes are biased, it is still far superior to many solid state high end designs. It takes many transistors in a circuit to come close to the amplification ability and linearity *before distortion reducing schemes* that it takes only one twin triode to do.

In an amplifier configuration using a 6NS7 tube (a twin triode) where one half of the tube amplifies and the other half acts as a follower, a biasing scheme where the output is taken from the cathode and it doesn't amplify voltage but current, but the first half does the amplifying of voltage. This particular circuit had an open loop (no NFB) distortion rating of less than 0.2%! Astonishingly low.

PUSH PULL AMPLIFIER

One of the controversies about push pull biasing is that

there exists a cross over distortion. What this means is that when the signal input to either tube causes the tube to cut off there is a less than linear transition for the other tube to "take up" when it begins to conduct for the other half of the wave (I hope you aren't confused *yet*).

In a pure class B biased circuit, this means that there is a gap between when one tube stops conducting and the other starts to conduct.

This makes for a wave form that kind of looks like this:

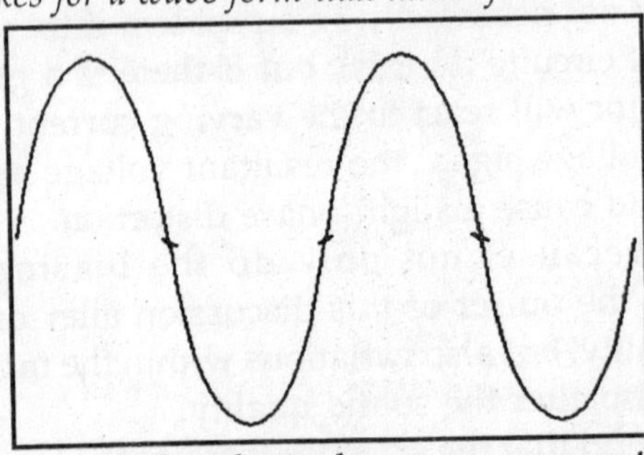

Notice the bends where the wave crosses the zero line. Class AB makes up for this by causing the tubes to conduct a little bit more of the cycle so as to overlap the cut off to turn on times. In other words, those bends are now in the signal past the zero crossing.

The result is a "purer" sine wave. The resulting spikes can then be filtered out with a small value capacitor that will not affect the audible high frequencies. Inverted feedback will also reduce this "crossover distortion". Also eliminated is the low volume level distorted sound of the crossover problem in pure class B operation. In essence, the tubes conduct as class A for low level signals, and as class B on loud level signals.

The real audible difference in a push-pull configuration, even in class A push-pull, is the fact of reduced even order harmonics. Or, increased odd order harmonics. Whichever textbook you wish to refer to, the fact is that odd order harmonics dominate in the push-pull stage. This is where the *real* sonic differences between push-pull and single ended amplifiers occur. This is not to imply that the crossover distortion is not real, just that the distortion is virtually eliminated by biasing and filtering. Transformers, too, naturally filter out the spikes. Solid state amps don't have transformers to do this, so many persons complain about the buzz, grainy or raspy sound they hear.

It is actually the crossover spikes they are listening to. It could also be a higher amplitude high end which causes the high end to sound (subjectively speaking) like an icing on cake, and not the cake itself, which is what we want to hear.

Better amps usually use a small choke in line with the output to the speaker to reduce the spikes that NFB doesn't. OTL (output transformer less) tube amps usually work in class A anyway so there is no problem with cross-over distortion there either.

In many high end tube amplifiers there are two potentiometers per channel that are user accessible. They are labeled "bias".

These potentiometers are there to set the bias for the output tubes so that they match each other. This allows both tubes to amplify their respective half of the signal *equally*. This also causes the tubes to cancel the even order harmonics (or *whatever*).

Many audiophiles mis-align the biasing. They claim that the amp sounds better. What is really happening is that now even order harmonics are showing up because of not being as canceled out by the push pull configuration.

MOSFET INTERNAL CAPACITANCES

The first component of capacitive parasitics we will examine is the MOSFET capacitances. These parasitic components are mainly responsible for the intrinsic delay of logic gates, and they can be modeled with fairly high accuracy for gate delay estimation. The extraction of transistor parasitics from physical structure (mask layout) is also fairly straight forward.

The parasitic capacitances associated with a MOSFET are lumped elements between the device terminals.

Based on their physical origins, the parasitic device capacitances can be classified into two major groups:

- Oxide-related capacitances.
- Junction capacitances.

The gate-oxide-related capacitances are Cgd (gate-to-drain capacitance), Cgs (gate-to-source capacitance), and Cgb (gate-to-substrate capacitance).

Notice that in reality, the gate-to-channel capacitance is distributed and voltage dependent. Consequently, all of the oxide-related capacitances described here change with the bias conditions of the transistor.

Figure below shows qualitatively the oxide-related capacitances during cut-off, linear-mode operation and saturation of the MOSFET. The simplified variation of the three capacitances with gate-to-source bias voltage is shown

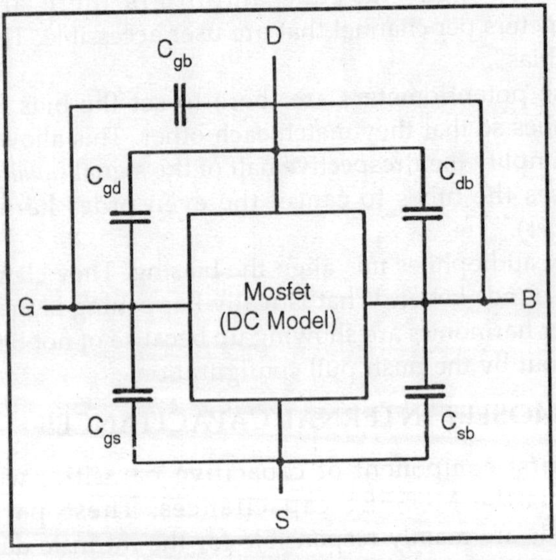

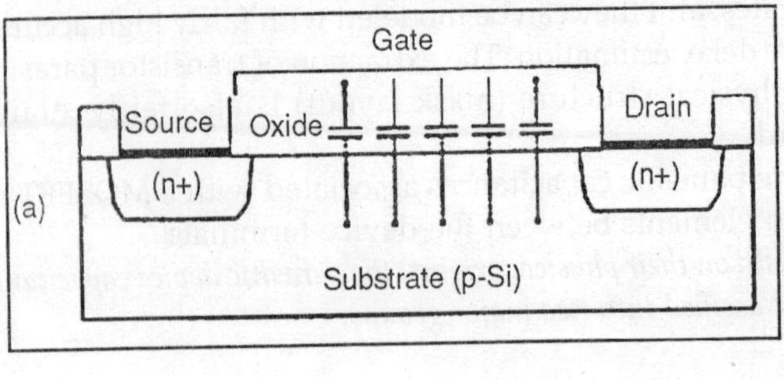

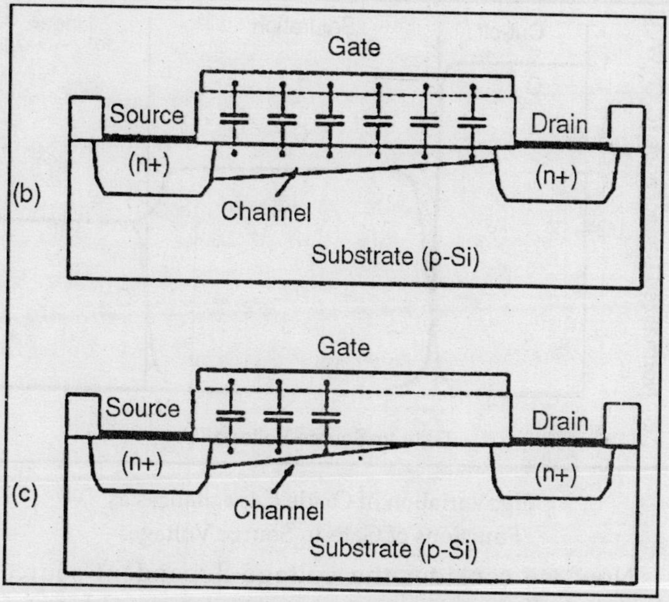

Fig: Schematic Representation of MOSFET Oxide Capacitances During (a) Cut-off, (b) Linear- mode Operation, and (c) Saturation.

We see that the total gate oxide capacitance is mainly determined by the parallel-plate capacitance between the polysilicon gate and the underlying structures.

Hence, the magnitude of the oxide-related capacitances is very closely related to:

- The gate oxide thickness,
- The area of the MOSFET gate.

Obviously, the total gate capacitance decreases with decreasing device dimensions (W and L), yet it increases with decreasing gate oxide thickness.

In sub-micron technologies, the horizontal dimensions (which dictate the gate area) are usually scaled down more easily than the horizontal dimensions, such as the gate oxide thickness.

Consequently, MOSFET transistors fabricated using sub-micron technologies have, in general, smaller gate capacitances.

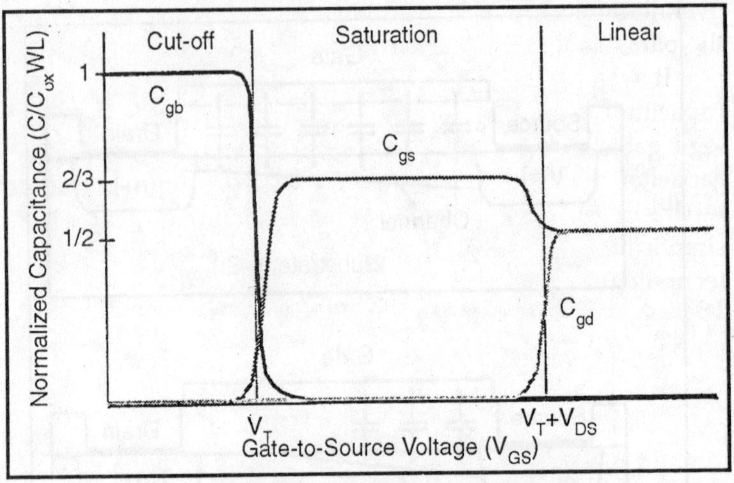

Fig: Variation of Oxide Capacitances as Functions of Gate-to-Source Voltage.

Now we consider the voltage-dependent source-to-substrate and drain-to-substrate capacitances, Csb and Cdb. Both of these capacitances are due to the depletion charge surrounding the respective source or drain regions of the transistor, which are embedded in the substrate.

Figure shows the simplified geometry of an n-type diffusion region within the p-type substrate. Here, the diffusion region has been approximated by a rectangular box, which consists of five planar pn-junctions.

The total junction capacitance is a function of the junction area (sum of all planar junction areas), the doping densities, and the applied terminal voltages.

Accurate methods for estimating the junction capacitances based on these data are readily available in the literature, therefore, a detailed discussion of capacitance calculations will not be presented here.

One important aspect of parasitic device junction capacitances is that the amount of capacitance is a linear function of the junction area. Consequently, the size of the drain or the source diffusion area dictates the amount of parasitic capacitance. In sub-micron technologies, where the overall dimensions of the individual devices are scaled down, the parasitic junction capacitances also decrease significantly.

It was already mentioned that the MOSFET parasitic capacitances are mainly responsible for the intrinsic delay of logic gates. We have seen that both the oxide-related parasitic capacitances and the junction capacitances tend to decrease with shrinking device dimensions, hence, the relative significance of intrinsic gate delay diminishes in sub-micron technologies.

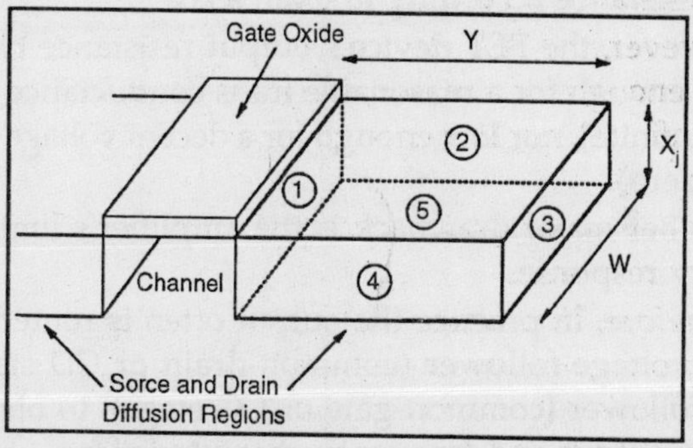

FREQUENCY RESPONSE OF CS AMPLIFIER

In a common-source amplifier, one of three basic single-stage field-effect transistor (FET) amplifier topologies, typically used as a voltage or trans conductance amplifier. The easiest way to tell if a FET is common source, common drain, or common gate is to examine where the signal enters, and leaves. The remaining terminal is what is known as "common". In this example, the signal enters the gate, and exits the drain. The only terminal remaining is the source.

This is a common-source FET circuit.

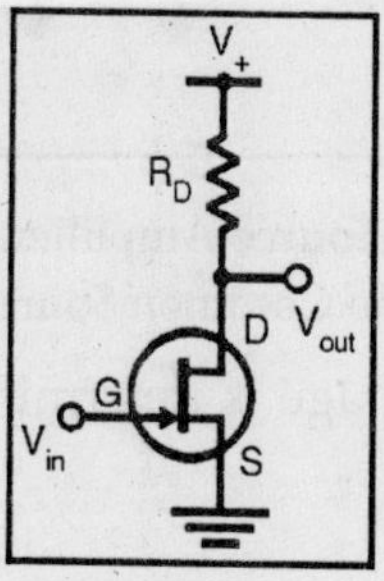

The analogous bipolar junction transistor circuit is the common-emitter amplifier.

·The common-source (CS) amplifier may be viewed as a trans conductance amplifier or as a voltage amplifier. (See classification of amplifiers).

As a trans conductance amplifier, the input voltage is seen as modulating the current going to the load. As a voltage amplifier, input voltage modulates the amount of current flowing through the FET, changing the voltage across the output resistance according to Ohm's law.

However, the FET device's output resistance typically is not high enough for a reasonable trans conductance amplifier (ideally infinite), nor low enough for a decent voltage amplifier (ideally zero).

Another major drawback is the amplifier's limited high-frequency response.

Therefore, in practice the output often is routed through either a voltage follower (common-drain or CD stage), or a current follower (common-gate or CG stage), to obtain more favorable output and frequency characteristics.

The CS–CG combination is called a cascode amplifier,

Consider a circuit:

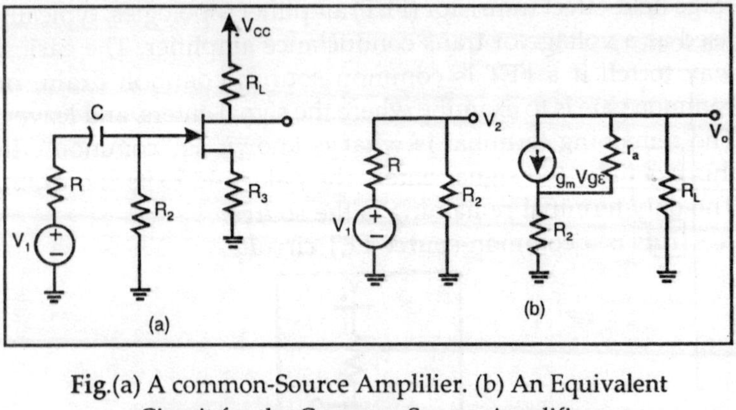

Fig.(a) A common-Source Amplilier. (b) An Equivalent Circuit for the Common Source Amplifier.

- The source voltage is determined from the following equations

$$\frac{V_D - V_2}{\gamma_d} + \frac{V_D}{R_1} + g_m V_{\varepsilon 2} = 0$$

$$and \frac{V_d}{R_2} + \frac{V_2 - V_0}{\gamma_d} = g_m V_{\varepsilon 2}$$

- Since the current leaving output node and source is zero
- The current through R_s The source voltage is

$$V_3 = \frac{V_0 - V_s}{R_L}$$

- The voltage gain is

$$\frac{V_0}{V_i} = \frac{-g_m R_L r_d}{R_L + r_d + R_5 (1 + g_m r_d)}$$

- If rd >> R_s and R_L we have the following relationship

$$A_V = \frac{-g_m R_L}{1 + g_m R_2}$$

- For gm R_s >> 1 we have

$$\frac{V_0}{V_i} = \frac{R_L}{R_2}$$

PAL ARCHITECTURE

The PAL architecture consists of two main components: a logic plane and output logic macrocells.

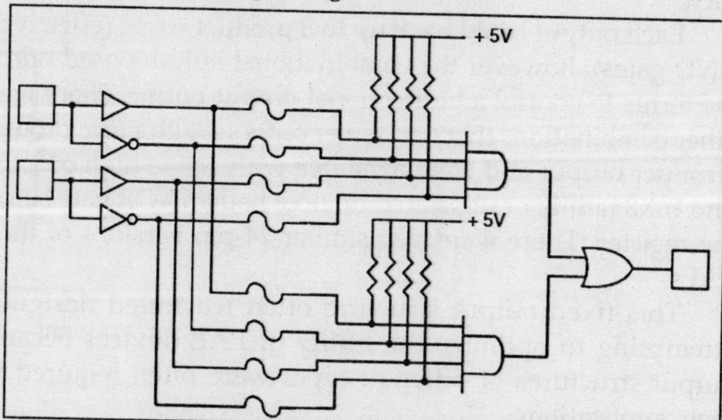

Fig. Simplified Programmable Logic Device

Programmable Logic Plane

The programmable logic plane is a programmable (PROM) array that allows the signals present on the devices pins (or the logical complements of those signals) to be routed to an output logic macrocell.

PAL devices have arrays of transistor cells arranged in a"fixed-OR, programmable-AND" plane used to implement"sum-of-products" binary logic equations for each of the outputs in terms of the inputs and either synchronous or asynchronous feedback from the outputs.

Output Logic

The early 20-pin PALs had 10 inputs and 8 outputs. The outputs were active low and could be registered or combinational.

Members of the PAL family were available with various output structures called"output logic macrocells" or OLMCs. Prior to the introduction of the"V" (for"variable") series, the types of OLMCs available in each PAL were fixed at the time of manufacture.

The PAL16L8 had 8 combinational outputs and the PAL16R8 had 8 registered outputs. The PAL16R6 had 6 registered and 2 combinational while the PAL16R4 had 4 of each.

Each output could have up to 8 product terms (effectively AND gates), however the combinational outputs used one of the terms to control a bidirectional output buffer. There were other combinations that had fewer outputs with more product term per output and were available with active high outputs. The 16X8 family or registered devices had an XOR gate before the register. There were also similar 24-pin versions of these PALs.

This fixed output structure often frustrated designers attempting to optimize the utility of PAL devices because output structures of different types were often required by their applications.

For example, one could not get 5 registered outputs with 3 active high combinational outputs. So, in June 1983 AMD introduced the 22V10, a 24 pin device with 10 output logic macrocells.

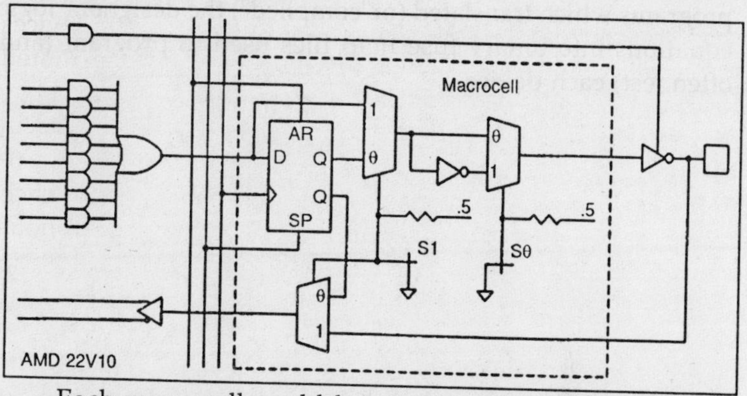

Each macrocell could be configured by the user to be combinational or registered, active high or active low. The number of product term allocated to an output varied from 8 to 16. This one device could replace all of the 24 pin fixed function PAL devices. Members of the PAL"V" ("variable") series included the PAL16V8, PAL20V8 and PAL22V10.

PROGRAMMING PALS

PALs were programmed electrically using binary patterns and a special electronic programming system available from either the manufacturer or a third-party, such as DATA/IO. In addition to single-unit device programmers, device feeders and gang programmers were often used when more than just a few PALs needed to be programmed.

For large volumes, electrical programming costs could be eliminated by having the manufacturer fabricate a custom metal mask used to program the customers' patterns at the time of manufacture; MMI used the term"hard array logic" (HAL) to refer to devices programmed in this way.

Programming Languages

Though some engineers programmed PAL devices by manually editing files containing the binary fuse pattern data, most opted to design their logic using a hardware description language(HDL) such as Data I/O's ABEL, Logical Devices'

CUPL, or MMI's PALASM. These were computer-assisted design (CAD) (now referred to as"design automation") programs which translated (or"compiled") the designers' logic equations into binary fuse map files used to program (and often test) each device.

Index

D

E

F

G

H